Radiocarbon Dating

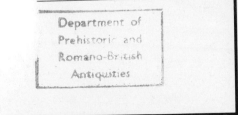

INTERPRETING·THE·PAST

RADIOCARBON DATING

Sheridan Bowman

Published for the Trustees of the British Museum
by British Museum Publications

© 1990 The Trustees of the British Museum
Published by British Museum Publications Ltd
46 Bloomsbury Street, London WC1B 3QQ

Designed by Andrew Shoolbred
Cover design by Slatter-Anderson

Set in Linotron Palatino and printed in
Great Britain by The Bath Press, Avon

British Library Cataloguing in Publication Data
Bowman, Sheridan
Radiocarbon dating. – (Interpreting the past)
1. Antiquities. Dating radiocarbon
I. Title   II. British Museum   III. Series
930.1′028′5

ISBN 0-7141-2047-2

Cover illustration: Section of the Belfast high-precision calibration curve for the recent past (courtesy of Gordon Pearson; see p. 46). The curve is superimposed on a section of a wooden timber which shows the annual growth rings of the tree. Tree rings are the basis of dendrochronological dating, which is used to provide the horizontal calendar-time axis of the calibration curve, here in years AD; the vertical axis is radiocarbon time BP (see ch. 4).

Contents

Acknowledgements

I particularly wish to thank Janet Ambers, Morven Leese, Wim Mook, Stuart Needham, Gordon Pearson, Jon Pilcher, Mike Tite and two kindly readers of the California University Press for reading all or parts of the text, and for making helpful comments (to say nothing of providing encouragement). However, any remaining gaffes are mine alone.

I am grateful to the following for providing illustrative material:

Cover illustration, fig. 5: Gordon Pearson

Figs. 2, 14, 21: The Trustees of the British Museum and Ian Longworth, Ian Stead and Jill Cook of the Department of Prehistoric and Romano-British Antiquities

Fig. 4: Laboratory of Tree-ring Research, University of Arizona

Fig. 7: Steve Dockerill

Fig. 8: The Trustees of the British Museum and Neil Stratford of the Department of Medieval Antiquities

Fig. 9: Lesley Fitton

Figs. 10, 12: The Trustees of the British Museum and the Department of Scientific Research

Fig. 11 (photo): The Research Laboratory for Archaeology and the History of Art, Oxford

Fig. 13: The British Society for the Turin Shroud

Fig. 16: Jon Pilcher

Fig. 17: Ulster Museum, Belfast

Figs. 18, 23, 26 (top): Minze Stuiver and Gordon Pearson

Fig. 22: The Somerset Levels Project

Figs. 24, 26: Stuart Needham

Preface

The primary aim of this book is to provide an introduction to the radiocarbon dating method, in particular in its application to archaeology. It starts from basic principles, but consideration is also given to the use and interpretation of radiocarbon results. The coverage is hopefully sufficiently wide to be of interest to a general audience and to archaeologists who might wish to use radiocarbon, but the text cannot be comprehensive.

In the format and space allowed, I have not been able to reference the many scientists and archaeologists whose work is summarised here. Their research fills not only many volumes of the journal *Radiocarbon*, but is also to be found in numerous other publications. Although I have been unable to cite them all by name, I wish to acknowledge and thank them here.

— 1 —

Background and Basic Principles

Carbon is a remarkable element: together with hydrogen, it is a component of all organic compounds and is fundamental to life. Both diamonds and graphite ('pencil lead') are pure carbon – only a rearrangement of the atoms distinguishes them. Diamonds, unfortunately, feature little in archaeology; instead it is the organic debris of past cultures which yields its secrets to radio-carbon dating.

Background

The existence of radiocarbon in nature was predicted before it was detected. Nevertheless, this prediction was sufficient for an American scientist called Willard Libby to perceive the basis of a dating method. The theoretical aspects were formulated in the mid 1940s when Libby was Professor of Chemistry at the University of Chicago. In 1946 he published a paper suggesting that radio-carbon might exist in living matter. One year later, a single-page paper appeared in the journal *Science* in which Ernest Anderson and Libby, together with collaborators in Pennsylvania, summarised the first detection of radiocarbon in material of biological origin. They showed that methane collected from the Baltimore sewage works had measurable radiocarbon activity, whereas methane manufactured from petroleum did not, and the implications of the findings for dating of carbonaceous materials were noted. These first experiments required enrichment of the radiocarbon in the sample to make it easily detec-table. By 1949, when Libby and Anderson (now joined in Chicago by James Arnold) published results of a world-wide assay of radiocarbon, enrichment was no longer necessary. The assay showed the contemporary level of radio-carbon in wood to be the same globally. The paper also contained the first two results of measurements on archaeological samples. The end of 1949 saw the publication of radiocarbon results on several samples of known age, and the publication of measurements of unknowns shortly followed. As Libby him-self later recalled, success was by no means a foregone conclusion: he and his colleagues persevered through a 'period of two or three years of secret research when we believed that the notion of radiocarbon dating was beyond reasonable credence'. A remarkable vision had been turned into an invaluable tool, and for his work on radiocarbon Libby was awarded the Nobel prize for chemistry in 1960.

Radiocarbon has had a major impact on archaeology, in particular on prehistory since the lack of a written record leaves much to conjecture. Previously all concepts of chronology were based on presumed linkages, however tenuous, with the civilisations of the Near East and Mediterranean, these being assumed in many cases to be the ultimate source of all innovation. For example, prior to the advent of radiocarbon dating, it was thought that monuments such as Stonehenge were later than the tholoi of Mycenae, on the assumption that megalithic monuments came from ideas that had diffused westwards and northwards. Incredulity undoubtedly greeted the radiocarbon results showing Stonehenge to be not only earlier than Mycenae, but older by several centuries once it was realised that the radiocarbon results required calibration to give calendar ages. Even in the historical period, dating techniques can play a major role. Not all written records are as detailed, precise or as durable as those of the Romans. In different cultures written records have been used for different purposes, so that chronological documentation of events can be more or less prominent depending on period and region.

Basic principles

Provided the reader has some familiarity with scientific terminology, the principles of radiocarbon dating can be fairly briefly stated and readily understood. However, in practice, various factors must be taken into account which can affect radiocarbon concentration in specific environments or organisms.

Carbon has three naturally occurring isotopes, that is, atoms of the same atomic number but different atomic weights. These are designated ^{12}C, ^{13}C and ^{14}C in scientific notation, the letter C being the symbol for elemental carbon and the isotopes having atomic weights 12, 13 and 14 respectively. They do not occur equally: carbon consists of 99% of ^{12}C, 1% of ^{13}C, but only about one part in a million million of modern carbon is ^{14}C. Unlike ^{12}C and ^{13}C, ^{14}C is unstable and therefore radioactive, though only weakly. Hence the name 'radiocarbon' for this isotope which, because of its scientific designation, is also called 'carbon fourteen'. In this book the notation '^{14}C' will be used when the isotope specifically is being referred to and the term 'radiocarbon' when discussing the dating technique in more general terms.

The really unusual characteristic of ^{14}C is that it is continually being formed. This occurs in the upper atmosphere (strictly the lower stratosphere and upper troposphere) by the interaction of neutrons produced by cosmic rays with nitrogen atoms. ^{14}C is therefore one of a small number of *cosmogenic* nuclides. After formation, the ^{14}C atoms rapidly combine with oxygen to form carbon dioxide which is chemically indistinguishable from carbon dioxide containing either of the other carbon isotopes. This carbon dioxide mixes throughout the atmosphere, dissolves in the oceans and, via the photosynthesis process and the food chain, enters all plant and animal life, known collectively as the biosphere. Under certain circumstances, in particular if the production rate is constant, there is a dynamic equilibrium between formation and decay, and therefore a constant ^{14}C concentration in the atmosphere. Thus in principle there is a constant ^{14}C level in all living organisms.

When a plant or animal dies, it ceases to participate in carbon exchange with the biosphere and no longer takes in ^{14}C. Were ^{14}C stable, its concentration would remain constant after death, but since it is not, the level falls at a rate that is determined by the law of radioactive decay. This law relates the number of atoms A left after time t to the initial number A_0 at time zero by the equation describing exponential decay:

$$A = A_0 \, e^{-\lambda t}$$

where λ is a constant equal to the reciprocal of the meanlife τ. A term better known in relation to radioactive decay is the half-life, $T_{1/2}$ (fig. 1). The half-life is related to the meanlife by

$$T_{1/2} = (\ln 2)\tau$$

or

$$T_{1/2} = 0.693\tau$$

where ln is the natural logarithm to the base e. Meanlives and therefore half-lives are specific to a particular radioactive atom, and for ^{14}C the best estimate of $T_{1/2}$ is 5730 years. For historical reasons, however, the Libby half-life is conventionally used in the calculation of a radiocarbon result (see p. 42). This 'half-life', being 5568 years, is 3% smaller than the true one and the corresponding meanlife is 8033 years. To determine the radiocarbon age, the equation at the top of this page is often written as

$$t = -\tau \, \ln(A/A_0)$$

or

$$t = -8033 \ln(A/A_0)$$

The constant percentage decrease of ^{14}C with time means that a small change in A results in a proportional change in the age t. Thus if the small change is a loss of 1%, t changes by 8033/100, an increase of about 80 radiocarbon years. This is a useful rule of thumb which will be used in some of the discussions

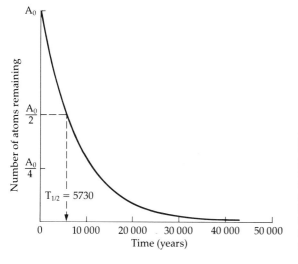

Number of atoms remaining — Time (years)

$T_{1/2} = 5730$

A_0, $\dfrac{A_0}{2}$, $\dfrac{A_0}{4}$

0   10 000   20 000   30 000   40 000   50 000

1 The decay of a radioactive element follows the exponential decay law. The primary characteristic of exponential decay is that the percentage decrease in number of atoms per unit time is constant; hence after each half-life the number of atoms remaining is halved: if there are A_0 atoms to begin with, then after one half-life there will be $A_0/2$ atoms remaining; after two half-lives, $A_0/4$ remain; after three, $A_0/8$ and so on.

that follow. The form of the equation is also such that an equivalent change in t would result if, for example, there were a 1% error in the value of A_0, but in this case a decrease of 1% causes a decrease of 80 radiocarbon years.

In principle, therefore, if the number of ^{14}C atoms remaining and the initial, or equilibrium, number can be evaluated by experiment, then the time elapsed since death can be determined. For a bone excavated on an archaeological site this provides an estimate of the time since death of the animal, though it must not be assumed necessarily to date the age of the context (for example, the stratigraphic layer) in which the bone was found.

There are two methods of measurement of ^{14}C. The so-called conventional method detects the activity of the sample, that is, the number of electrons emitted per unit time and weight of sample by the decay of ^{14}C. The other method, accelerator mass spectrometry (AMS), is a very much more recent technique and directly detects the number, or a proportion of the number, of ^{14}C atoms in the sample relative to ^{13}C or ^{12}C atoms. Both perform similar measurements on modern reference standards to establish the initial activity or concentration ratio of ^{14}C. These techniques are discussed in more detail in chapter 3.

Datable materials

In general, the materials which can be dated by radiocarbon are those which once formed part of the biosphere and are therefore organic. For example, the most commonly preserved sample types occurring on British sites are bone, shell and charcoal, but on some sites or in other areas of the world a different suite of materials might remain. Preservation may be effected by charring, as

with charcoal, so that dating of charcoal inclusions in wrought iron and of food residues on, and organic temper in, pottery is feasible. Equally, uncharred wood and other plant remains such as ropes, cloth, reeds and seeds may be well preserved in arid environments or if waterlogged. Certain situations will be detrimental to some types of material. Peat bogs, themselves datable, are acidic and will dissolve bone and shell. Ironically, though, they preserve muscle and other soft tissue otherwise only found in arid environments, such as existed in many of the pyramids. Soft-tissue remains are also datable by radiocarbon, as are many other materials such as antler, horn, tooth, ivory, hair, blood residues, wool, silk, leather, paper, parchment, insects and coral (fig. 2). Sediments and soils may also be datable, although the sources of carbon within these are many and diverse and such material is rarely dated for archaeological purposes.

In some circumstances it may even be possible to date materials that have not been part of the biosphere, if their formation involves incorporation of carbon with a ^{14}C concentration that can be assumed to be in equilibrium with the atmosphere. Mortar, for example, involves slaking of lime when atmospheric carbon dioxide is absorbed on hardening, but there are many complicating factors.

Not all materials from all situations are datable with the same degree of success, and some of the problems will be outlined in later sections.

The global carbon cycle

The atmosphere, oceans and biosphere are reservoirs of carbon and are part of the global carbon cycle shown in figure 3. To understand some of the factors

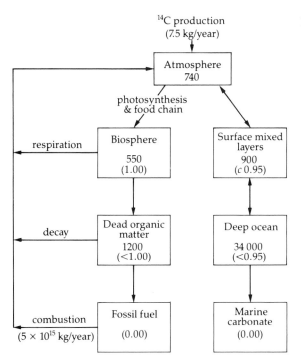

^{14}C production
(7.5 kg/year)

2 (*opposite*) Materials datable by radiocarbon are mainly organic, such as these antlers of red deer (*Cervus elephas*) found in the galleries of the Neolithic flint mines known as Grimes Graves (at Weeting, near Thetford in Norfolk, England). Flint forms bands in chalk and Neolithic miners used the antlers to prise out chalk blocks. The flint thus removed would then have been knapped to make a variety of implements. These flint formations were exploited in the Late Neolithic, *c.* 4500 years ago.

3 The atmosphere, oceans and biosphere are reservoirs of carbon, along with other reservoirs, as shown in this simplified global carbon cycle. The approximate carbon content of each reservoir, where known, is given in units of 10^{12} kg (million million kg). The atmosphere contains only about 2% of the exchangeable global carbon. The relative activities of ^{14}C in each reservoir are given in brackets.

that affect ^{14}C concentration in specific environments, the relative sizes of and interactions between the reservoirs need to be considered. The main exchanges and transfers of interest here are the uptake already mentioned of atmospheric carbon dioxide by the biosphere, and the exchange process between the atmosphere and ocean or other surface waters such as lakes. It can be seen from figure 3 that the ocean is in fact best considered as two parts: a mixing layer and the deep ocean. The situation in the oceans is quite complex and is discussed further in chapter 2.

Assumptions made in the simplified approach

The usefulness of any dating technique requires that it be applicable to the materials commonly found on archaeological sites and have a sufficiently low error term to allow temporal differentiation. Equally fundamental is the need for the method to be globally applicable; it should also be valid at all, or a good range of, periods in the past. In the case of radiocarbon this in turn requires a global level of ^{14}C in the atmosphere that has not changed with time; in addition, the biosphere should be in equilibrium with the atmosphere. Theoretically these equilibria exist under certain circumstances, and at this stage it is perhaps appropriate to review some of the assumptions made, explicitly or implicitly, in setting out the basis of the radiocarbon dating method:

- The atmosphere has had the same ^{14}C concentration in the past as now; this in turn assumes constant production, constant and rapid mixing, exchange and transfer rates, as well as constant sizes of reservoirs.
- As a corollary of this, the biosphere has the same overall concentration as the atmosphere and therefore it is assumed that there is rapid mixing between these two reservoirs.
- The same ^{14}C concentration exists in all parts of the biosphere.
- The death of a plant or animal is the point at which it ceases to exchange with the environment.
- After ceasing exchange, the ^{14}C concentration in a plant or animal is only affected by radioactive decay.

None of these assumptions is strictly correct, beyond a rough first approximation. Much of the following chapters will be spent discussing the geochemical and geophysical reasons for the breakdown of these assumptions and the ways in which these problems are dealt with. They can be summarised briefly as:

- processes affecting the global concentration of ^{14}C in the atmosphere
- source or reservoir effects
- alteration effects
- contamination.

Processes affecting the global concentration are largely production-rate variations, but under some circumstances the size of the atmospheric carbon reservoir has been changed. Source or reservoir effects result from the particular origin of the carbon taken up by an organism; this may have local effects dependent on the prevailing ^{14}C to ^{12}C balance. Alteration effects is the term used here to describe processes, other than radioactive decay, that change the concentration of ^{14}C in an organism relative to that of the atmosphere or other parts of the biosphere. Strictly speaking, contamination could be considered under this heading, but this alters the apparent ^{14}C concentration in the sample rather than the true one by introducing extraneous carbon material.

Age at death and time-width of samples

The processes outlined above all potentially affect the ^{14}C concentration in a sample submitted for dating and are discussed in chapter 2. The one remaining assumption is that time of death and cessation of exchange with the biosphere are contemporary events. If not, then the radiocarbon age of the organism at death is not zero. This is one type of 'age offset'. Others arise from reservoir effects and archaeological depositional processes, and their implications for dating archaeological sites are considered in chapter 5.

The time-width of a sample is the total growth and exchange period represented. Measurement of a sample with a significant time-width gives an average ^{14}C activity that depends on the relative proportions of the components present. The time-width also affects the way the radiocarbon result is converted to a calendar age (see ch. 4).

For seeds and grasses, since only a single season of growth is involved, there is no inherent age offset. Bone does not cease to exchange with the biosphere until death, but there is a turnover time of about thirty years for human bone and an equivalently shorter period for animal bone. Hence there is no age offset, but there is a time-width for bone samples.

The outstanding example of age at 'death', or more usually felling, is wood. It is well known that trees grow by the addition of rings, usually though not always annually. Once laid down, rings cease to exchange with the biosphere. Hence, if one considers a long-lived tree, say a three-hundred-year-old oak, the innermost heartwood will give a radiocarbon result 300 years older than the sapwood. Indeed, this is as it should be. However, if part of that heartwood were found on an archaeological site, the radiocarbon result would not provide the date of usage of the wood, but rather a date 300 radiocarbon years earlier; more had it been seasoned before use or re-used. This is the 'old wood' problem.

The time-width for wood depends on the number of rings taken for dating. If the sample is fragments of charcoal, the overall time-width may be very variable and is unquantifiable.

— 2 —

Radiocarbon Concentration Effects

Many factors affect the concentration of ^{14}C in plants and animals before their death and others, including radioactive decay, affect these levels even after death. These effects, introduced in chapter 1, are discussed more fully in the following sections.

Atmospheric ^{14}C variations

The work of the early decades
The assumption of constant ^{14}C concentration was not taken lightly even from the very beginning. In 1949, James Arnold and Willard Libby published a 'curve of knowns' which was a test of the technique, and therefore of the assumption of constant concentration, using known-age samples ranging from about 900 to 4900 years old. Given the experimental conditions then achievable, there was good agreement, at least for this period, between the theoretical and measured ^{14}C activities versus known age. During the 1950s, with advances in techniques for detecting ^{14}C, discrepancies increasingly emerged between radiocarbon ages and historical ages for the Egyptian Old Kingdom. These discrepancies were far from insignificant, the radiocarbon results being several centuries too young. The validity of the historical ages was of course not proven beyond doubt, and other evidence was sought.

Tree rings provided the truly known-age material needed to test the accuracy of the new technique. Dendrochronology, the science of using tree rings for dating (see ch. 4), had been developed by A. E. Douglass in America in the early part of the twentieth century for research on past climate. By the late 1950s several scientists, notably Hessel de Vries in the Netherlands, were radiocarbon dating rings from trees dated by dendrochronology, and confirming the radiocarbon discrepancy. It therefore became clear that radiocarbon results would need to be calibrated to convert them to calendar ages. Since there is no theoretical way of predicting the correction factor, empirical calibration curves were needed to link radiocarbon 'age' with known age.

In the 1960s, a continuous tree-ring sequence stretching back some 8000 years was established by Wesley Ferguson, and the first calibration curve using

4 The bristlecone pine (*Pinus aristata*) grows in the White Mountains of California at an altitude of about 3 km. This species can live for more than 4000 years, adding a ring per year; however, due to the short growth season, these rings are very narrow and even a long-lived specimen may have a diameter of little more than 2 m. Dead trees survive *in situ* because they do not suffer decay owing to their high resin content and the dry environment. Both living and dead bristlecone pines were therefore used to establish a dendrochronological sequence some 8000 years long. The living trees did not need to be felled: the cross-sections required for study were obtained by boring into the trunks.

this was published by Hans Suess. This curve was partly based on a remarkable tree, the bristlecone pine (fig. 4).

Suess's curve confirmed that there are indeed major discrepancies between radiocarbon age and calendar age. This was the first useful calibration curve in that it had a long temporal coverage, radiocarbon error terms in the order of 1%, and used truly known-age material (i.e. tree rings) for the calendar axis. It was also the first of many such curves, and their proliferation prior to 1985 has caused almost as many problems as have been solved. There are now internationally agreed calibration curves for the period back to 2500 BC, and the use of these is discussed in chapter 4.

Two trends were apparent in Suess's curve. First there is a long-term trend that can be described approximately by a sine wave with a period of about 9000 years. The maximum deviation from true age is about 900 years too recent at the beginning of the fourth millennium BC. On the other hand, in the middle of the first millennium AD, radiocarbon produces ages too old by a century or so (fig. 5).

The second feature takes the form of 'wiggles'. These are superimposed on the main sine wave and are of short calendar duration (a few decades) but can have amplitudes on the radiocarbon axis of a century or so. Suess,

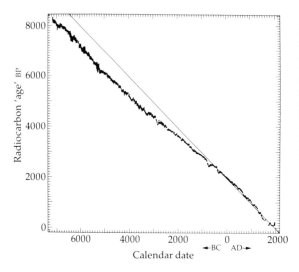

5 High-precision radiocarbon calibration curve for the past 7000 years based on Irish oak (*Quercus* sp.). The straight line is the ideal 1:1 correspondence between radiocarbon age and dendrochronological (calendar) age. In the true curve, the general sinusoidal trend is very clear, with radiocarbon ages consistently too young in the BC period. Superimposed on this are wiggles of short calendar duration (a few decades) but which can have amplitudes on the radiocarbon axis of a century or so (see fig. 18).

when asked about the line drawn through his data points, said he had used the process of 'cosmic schwung', what we might call 'freehand'. Whether these wiggles really existed or were an artefact of experimental error was the subject of much investigation in the 1970s. Techniques with higher precision have now been developed and the general validity of these wiggles has been proven by various laboratories, notably those in Belfast, Groningen, Heidelberg and Seattle.

Variations in natural production rate
Although the detailed geophysical causes of the trends may not be fully understood, the broad principles are known. The long-term variation correlates reasonably well with fluctuations in the earth's magnetic field strength (the geomagnetic moment). The geomagnetic moment affects ^{14}C production because cosmic rays are charged particles and are therefore deflected by a magnetic field. If the magnetic moment is high, more cosmic rays are deflected away from the earth and production of ^{14}C will fall; if low, the production rises. Whether the moment is high or low, the effect of the magnetic field varies with latitude, but rapid mixing in the atmosphere leads to a uniform ^{14}C concentration globally. However, when the production rate changes, a new equilibrium concentration in the carbon cycle as a whole (fig. 3) will only be established after a considerable time, owing to the finite, and in some cases long, mixing and exchange rates and the relative sizes of the different reservoirs. The likely timescale for achieving the new equilibrium level throughout is of the order of 10 000 years, although the atmosphere, biosphere and surface oceans require only a few tens of years to adjust to quasi-equilibrium. Quantification of the effect that these natural production rate changes have on ^{14}C concentration is impossible primarily because the sizes of the fluctuations themselves are unknown.

Not only can the strength of the earth's magnetic field change, but in the past the direction of the field is known to have reversed relative to that of today. If global, true reversals and polarity excursions (more rapid, often local, reversals) would have affected ^{14}C production significantly due to the low value

18

of field strength associated with the transition from one field direction to the other. Such effects can be ruled out because there have been no reversals nor unequivocally global excursions in the timescale of radiocarbon.

So what causes the short-term variations seen in the calibration curve: the wiggles? These are known as the de Vries effect or Suess wiggles and are probably produced by variations in sunspot activity, records for which over the past few centuries show a cycle with period of about 200 years, superimposed on which is a more rapid 11-year cycle. High sunspot activity increases the weak magnetic field that exists between the planets, and at such times there is greater deflection of cosmic rays and hence ^{14}C production decreases.

It was previously thought that the 11-year cycle was likely to produce major difficulties if samples of a single year's growth (e.g. grasses) were dated. This was on the assumption of a large shift in atmospheric concentration year-to-year. Several workers have attempted to quantify the likely effect using consecutive single tree rings and, more interestingly, samples of single-year growth such as vintage wines or malt whiskies! Of course, great care has to be taken to reduce experimental error, otherwise small fluctuations could be masked. It now seems that the effect of the short cycle is unlikely to cause more than about 20 years' variation in age.

The 200-year cycle does, however, have a significant legacy. The wiggles associated with this can represent changes in radiocarbon age of a century or two when the corresponding calendar age has changed by only a few decades. It is these wiggles, more than the long-term trend, which provide problems for converting ^{14}C results to true calendar ages (ch. 4).

Natural changes due to glaciations
The solubility of carbon dioxide in water depends on temperature, increasing as temperature drops. Past glacial periods will therefore have had a significant effect on the amount of ^{14}C in the atmosphere. Warm interglacial periods would not only reverse this trend, but would also release aged carbon (carbon depleted in ^{14}C) from the ice masses formed during the glaciation. It would also have encouraged plant growth and therefore an increase in the animal population. The effect of these interrelated factors on radiocarbon dates has not yet been established.

The effect of recent human activity on atmospheric ^{14}C content
Not only has nature been fickle in producing ^{14}C variably with time, but in two ways man has also had an effect on the global level of ^{14}C. The first is the fossil-fuel or Suess effect. It was recognised that tree rings corresponding to the early half of the twentieth century had a significantly lower ^{14}C content than expected, and this was shown to be due to the burning of fuel such as coal (see fig. 6).

This has had one major implication for the practice of radiocarbon dating: no recent organic material can be used as a modern standard. Instead, oxalic acid stocked by the US National Bureau of Standards has been adopted and its measured ^{14}C activity related to that theoretically predicted, in the absence of the fossil-fuel effect, for a wood sample grown in AD 1950. The use of this standard and the year AD 1950 as the zero point of the radiocarbon timescale are part of the convention for quoting radiocarbon results (see p. 42).

A more dramatic effect on atmospheric ^{14}C content has come about through

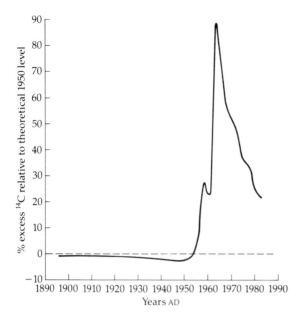

Years AD

6 Human influence on global ^{14}C levels can be seen in the fossil-fuel and bomb effects. The burning of large quantities of fossil fuel such as coal started in the last century. Coal is fossilised plant material laid down so long ago that its ^{14}C has totally decayed away. When it is burnt in large quantities, the carbon dioxide released dilutes the atmospheric ^{14}C concentration relative to ^{13}C and ^{12}C and changes both the size and isotopic composition of the atmospheric carbon reservoir. The ^{13}C and ^{12}C released is distributed only slowly through the whole carbon cycle. In contrast to the fossil-fuel or Suess effect, nuclear-weapons testing has produced large quantities of ^{14}C. The result of testing in the 1950s and 1960s was roughly to double the atmospheric ^{14}C content as measured in about 1965. The banning of atmospheric testing and the gradual mixing of the excess ^{14}C through the carbon exchange cycle has decreased this level, so that currently it is about 20% higher than the theoretical 1950 level as measured by the activity of the oxalic acid reference standard.

nuclear-weapons testing (fig. 6). This is often referred to as the bomb effect. The neutrons produced in turn produce ^{14}C by interaction with ^{14}N, simulating the natural cosmogenic production, albeit in large bursts. Using this artificial injection of ^{14}C to good purpose, radiocarbon scientists have been able to test theories about the mixing rates of ^{14}C through the various carbon reservoirs. They have also shown that once laid down, a tree ring does not exchange ^{14}C with other rings; this is fundamental to the use of dendrochronology to construct a radiocarbon calibration curve.

Alteration effects

This term is used here to encompass the effects that may change the ^{14}C concentration in a sample, making it either different from that in the atmosphere or different from the value expected purely on the basis of decay. The processes involved are fractionation, recrystallisation of shell carbonate and *in situ* production. They potentially alter the ^{14}C concentration of the true sample material without any infiltration of extraneous carbon-containing material. Contamination is discussed separately below. Of the alteration effects listed, fractionation is the most important; it applies to every sample dated.

Fractionation
Although ^{12}C, ^{13}C and ^{14}C are all carbon isotopes and chemically indistinguishable, in any biological pathway there will be a tendency for the lightest isotope ^{12}C to be preferentially taken up. Similarly ^{13}C will be taken up in preference to ^{14}C. Growing plants and animals (that is, the parts still exchanging with the biosphere) are therefore expected to have a lower ^{14}C level than the atmosphere. If the difference is significant, they will appear to be older than the atmosphere when dated and, since uptake also varies according to species, different parts of the biosphere will appear to have different radiocarbon ages.

This differential uptake is referred to as fractionation, and needs to be taken into account if useful radiocarbon results are to emerge. Fortunately, the fact that carbon has three isotopes of which two are stable enables a correction for fractionation to be applied. The principle is to measure the ratio of ^{13}C to ^{12}C in the sample; any shifts in this concentration ratio, sample to sample, indicate fractionation has occurred. To evaluate what the equivalent shifts in ^{14}C concentration ratio would be, it is assumed that the effect for ^{14}C is double that for ^{13}C, reflecting the mass difference relative to ^{12}C. To quantify the effect, the ^{13}C concentration is measured relative to a standard referred to as PDB, a Cretaceous belemnite (*Belemnitella americana*) from the Peedee formation of South Carolina which was first used as the standard.

The ratio of ^{13}C to ^{12}C can be readily measured in a mass spectrometer with low resolution, since ^{13}C is far more abundant than ^{14}C (1% relative to one part in a million million of ^{14}C in modern carbon). In accelerator mass spectrometry (AMS) dating (see ch. 3), the ratio can be measured as part of the measurement process. In conventional dating a sample of the carbon dioxide is taken after combustion of the sample (or after acid dissolution if the sample is a carbonate).

The ^{13}C concentration ratio in the sample can then be expressed as a $\delta^{13}C$ value, where

$$\delta^{13}C = \left[\frac{(^{13}C/^{12}C)}{(^{13}C/^{12}C)_{PDB}} - 1 \right] \times 10^3 \, \%o$$

As the symbol %, or percent, indicates parts per hundred, so ‰ (referred to as permil) indicates parts per thousand. As well as measurement relative to a standard, there must be an agreed value to which the $\delta^{13}C$ is corrected; for radiocarbon dating this value is $-25‰$. This is approximately the value for wood, though any other value would do equally well provided it were universally used. The fractionation corrected ^{14}C activity of the sample (A_c) relative to the measured activity (A_m) is given by

$$\frac{A_c}{A_m} = \left[\frac{1 + (-25/10^3)}{1 + (\delta^{13}C/10^3)} \right]^2$$

This rather daunting expression can be simplified to give an approximate age difference:

$$t_c - t_m \approx 16(\delta^{13}C + 25) \text{ years}$$

This means an age correction of about 16 years for every 1‰ difference from $-25‰$. If the $\delta^{13}C$ value is larger than $-25‰$, the corrected age is larger (older) than the measured age (for example, if $\delta^{13}C$ were $-15‰$, older by about 160 radiocarbon years).

How significant this effect can be is shown in table 1, where a typical range of $\delta^{13}C$ values is listed. In particular, marine carbonates have $\delta^{13}C$ values in the region of 0‰. The fact that fractionation gives a higher ^{14}C content in ocean waters relative to terrestrial plant life is rather ironic since the mixing and upwelling effects for ocean waters, discussed more fully below, roughly compensate the effect. Since a $\delta^{13}C$ correction must be made for comparison of radiocarbon results in the rest of the biosphere, it must be used throughout, and a corrected radiocarbon age for a marine carbonate will thus appear to be 400 years too old relative to, say, contemporary wood.

7   The winter diet of the sheep of North Ronaldsay is dominated by seaweed, the most favoured of which are the brown kelps. These are plentiful in the winter following storms that tear them from the seabed and deposit them on the beaches. The sheep are largely excluded from pasture land by a wall, 1.8 m high and 19 km long, which rings the island. The sheep have developed a unique gut flora to cope with the high iodine content of their diet. The $\delta^{13}C$ values of bone and wool from these animals is in the region of $-13\%o$, substantially higher than those for normal grass-fed sheep (apparently the taste is rather different, too!). However, there is as yet no evidence that this food resource was similarly exploited in early settlement of the Orkney Islands. Despite finds of carbonised seaweed in archaeological deposits of both the Norse and Neolithic periods on Sanday, none of the animal bones indicate other than a terrestrial diet.

The atmospheric $\delta^{13}C$ value lies between the values for the biosphere and the oceans. This shows that the fractionation occurring in photosynthesis pathways tends to deplete the ^{14}C level relative to ^{12}C, but in the phase transition to the ocean there is an enrichment. It is important to note also that different photosynthesis pathways exist that cause very different levels of fractionation. The normal one is that for so-called C3 plants, the Calvin pathway. The Hatch–Slack pathway of C4 plants, such as maize, sugar cane, and grasses living in semi-arid conditions, gives $\delta^{13}C$ values which are larger (less negative) than those for C3 plants.

Animals reflect the $\delta^{13}C$ value of their food, though the actual value will be modified by the animal's own biological processes. Any animal that eats a predominantly marine diet (fig. 7) or subsists to a large degree on C4 plants will to some extent have a $\delta^{13}C$ value greater than an animal subsisting on C3 plants.

Table 1 *Approximate* $\delta^{13}C$ values for various materials*

Material	$\delta^{13}C$ value
Wood, peat and many C_3 plants	-25%o
Bone collagen	-19%o
Freshwater plants	-16%o
Arid zone grasses	-13%o
Marine plants	-12%o
Maize	-10%o
Atmospheric CO_2	-8%o
Marine carbonates	0%o

* The ranges on these data are typically ± 2 or 3%o but substantially more variability is possible. At 16 years per $\%$o age difference from -25%o, these data illustrate the need for fractionation corrections to measured radiocarbon results.

It is part of the convention for calculating a radiocarbon result to apply a fractionation correction, whether measured or assumed, and it is important to check that this has been done, particularly with dates published some time ago. It is preferable to measure the $\delta^{13}C$ value relevant to a particular sample wherever this is possible, rather than using an assumed average value for a species.

In situ production

The earth's atmosphere acts as a radiation shield so that the higher the altitude, the higher the cosmic ray flux. For example, at an altitude of 3 km, the flux is an order of magnitude higher than at sea-level, though still only 3% of that in the stratosphere, where most ^{14}C production occurs. The bristlecone pines used for the first calibration curve grow at such an altitude and, as wood contains a few percent of nitrogen, the possibility of *in situ* production of ^{14}C needed to be considered. Another mode of *in situ* production might be through neutrons produced in the electrical discharge associated with a flash of lightning. Bristlecone pines growing at altitude might be expected to be struck by lightning more frequently than low-altitude trees.

Such effects would of course be particularly pertinent the older the true age of the sample, since there would have been more time for *in situ* production, and hence there would be more discrepancy in the radiocarbon age with time. The overall trend in the bristlecone calibration showed just this tendency.

To attempt to simulate *in situ* production, wood samples have been exposed to high neutron fluxes in reactors, but without observable increase in ^{14}C activity. These findings are confirmed by the broad agreement between the bristlecone calibration and the high precision calibration curves (ch. 4) which are largely based on low-altitude trees.

Recrystallisation

For completeness, recrystallisation of shell carbonate is only mentioned here. Although recrystallisation and isotopic exchange can occur without actual chemical exchange with the environment of the sample, they are nevertheless intimately connected with the possibility of chemical exchange and are discussed as part of the section on contamination.

Source or reservoir effects

Under this heading are grouped the age shifts that are not accurately quantifiable and arise from the local environment of the organism assimilating carbon. Three major effects will be considered: marine, hard water and volcanic.

Marine

Although there is rapid global mixing in the atmosphere and terrestrial biosphere, mixing rates in the deep oceans are slow, so that radioactive decay becomes an important factor in the mixing between incoming carbon dioxide from the surface layers and outgoing from the deep layers. Hence really deep ocean waters of the present day can show a radiocarbon age of a few millennia. However, the system of mixing is not straightforward. In particular, deep waters can move upwards. This phenomenon is known as upwelling and is latitude dependent. It occurs predominantly in the equatorial region as a consequence of the trade winds, though various factors can cause local upwelling such as coastline shape, local climate and wind, as well as ocean bottom topography. Hence, although the time taken for equilibration of $^{14}CO_2$ in surface waters is of the order of 10 years, the degree of equilibration of the deep waters is not known. The upwelling of ^{14}C-depleted deep water means that the surface water has an apparent radiocarbon age relative to the atmosphere. This amounts

8 This carved whalebone plaque is purported to be of medieval Spanish origin and was radiocarbon dated to 1480 ± 80 BP (OxA-1164; see p. 42 for an explanation of the convention for quoting radiocarbon results). However, the marine effect means that at death the whale itself would have had an apparent age of several centuries. The true age of the object cannot therefore be accurately assessed, though the radiocarbon result is sufficient to demonstrate that it is of some antiquity rather than modern.

to *about* 400 years if the $\delta^{13}C$ values are normalised to $-25‰$. For marine carbonates, such as shells and corals, the measured $\delta^{13}C$ values are about 0‰; if no fractionation correction were made to the radiocarbon result, the effects of fractionation and the surface marine effect would approximately cancel. However, whereas the effects of fractionation can be accurately quantified, the marine effect cannot.

There are generalised measurements of the marine effect for broad oceanographic regions, but the difficulty is that local effects can predominate and indeed be very variable over relatively short distances. One way in which radiocarbon workers attempt to quantify these effects is by assuming there has been no change with time, and dating known-age shells of the same species from the same locality. These are generally relatively recent specimens, but collected before the nuclear weapons testing of the 1950s and 1960s. Such material has demonstrated that apparent ages differing by a few centuries can be obtained for localities in relatively close proximity. Of course, whether or not such a discrepancy has a profound effect on interpretation depends on the dating application: for an archaeologist requiring a date for a shell midden a possible systematic deviation of this magnitude might render the results useless, but for an oceanographer studying coastline changes it might be perfectly adequate.

It is not only marine carbonates that exhibit this reservoir effect; marine mammals such as whales and seals show an apparent radiocarbon age of several centuries (fig. 8).

Northern-to-southern-hemisphere effect
While there is good atmospheric mixing within each hemisphere, mixing between them is poor because their respective prevailing winds blow in opposite directions along the equator. There is evidence that radiocarbon results for the southern hemisphere are systematically about 30 radiocarbon years older than those for the northern hemisphere. The data come from calibration curves, with dendrochronology providing the absolute timescale in each case. The greater ocean surface area in the southern hemisphere is believed to be the cause; a degree of dilution of atmospheric ^{14}C occurs due to the greater interface between the atmosphere and the ^{14}C-depleted surface oceans. How this affects the equatorial region is not known. Long tree-ring sequences have not yet been developed and no radiocarbon measurements have been made.

Island effect
In the same way that an increased water mass causes the northern-to-southern age difference, it has been suggested that there might be an island effect. However, the strong mixing within each hemisphere argues against it, as do the high-precision calibration curves of the Belfast and Seattle laboratories, which are in agreement over a period some 4500 years long to within a few years (see ch. 4). If an island effect existed, the Belfast radiocarbon results on oak grown in Ireland would be affected, but the Seattle results on various species from the North American continent would not.

Hard-water effect
Although freshwater shells escape the ocean reservoir effects experienced by marine shells, they can suffer another effect: that of hard water. Worse still, the hard-water effect can also affect marine carbonates if they are deposited

in certain environments. The hard-water effect is so called because it is often associated with the presence of calcium ions resulting from dissolution of infinite-age calcium carbonate. However, there can be sources of carbon other than calcium carbonate, such as soil humic material, soil carbon dioxide and atmospheric carbon dioxide. Furthermore, the activity of the ^{14}C will depend not only on the source of carbon, but on the time elapsed between the carbon uptake by the water and its uptake by a plant or animal. Thus the presence of hardness (calcium ions) coincides with depleted ^{14}C concentration, but the size of the reservoir effect is not directly correlated with the amount of hardness. It affects living organisms such as molluscs and aquatic plants, can account for discrepancies of several centuries and is observed not only in fresh water, but also in marine environments where a substantial carbonate-rich freshwater influx is encountered (e.g. river mouths). The term is also applied to the age offset observed for terrestrial shells, for example snail shell, where the organism has been feeding in carbonate-rich areas such as chalkland. However, there is little evidence to suggest that it is a problem for terrestrial plants growing in hard-water areas. Here the carbon uptake appears to be dominated by the photosynthesis process.

The hard-water effect is not quantifiable since it is dependent on local factors; there cannot even be a general geographical guide to the likely age offset as there is for surface ocean waters. The approach taken is similar to that for dating of marine carbonate: assume no change with time, and evaluate the age offset using recent specimens of the same species from the same locality. Naturally, it is not always possible to locate appropriate pre-bomb specimens.

Volcanoes
Volcanoes, whether active or apparently dormant, issue gases including carbon dioxide. This carbon dioxide, coming from deep within the earth's crust, has no ^{14}C activity and therefore locally dilutes the atmospheric concentration. Plants growing in the vicinity of some volcanic vents have been shown to have

9 Ceramic storage vessels found at the Minoan town of Akrotiri, which was destroyed by the volcanic eruption of the Aegean island of Thera (Santorini). Dating of this destruction by radiocarbon and ceramic typology is subject to some debate, and other dating evidence for the event is only inferentially linked to the Thera eruption. As a massive eruption it would have sent large clouds of dust into the air, causing global cooling of the climate and in turn affecting tree growth. Frost-damaged tree rings have been observed in the American dendrochronological record at about the right time, as have narrow rings in the Irish oak sequence. Thera could also have injected high acidity levels into the atmosphere that are now seen recorded in ice cores. If the link between Thera and the tree-ring and ice-core data is proven beyond doubt, then the eruption would date to the second half of the seventeenth century BC and would indicate that, at least in this case, there was little or no volcanic effect on the radiocarbon results.

high apparent ages; for example, some on the island of Santorini in the Aegean show a millennium or more offset. Santorini is the modern name for Thera, on which stood the Minoan town of Akrotiri. The effect of volcanic carbon dioxide has caused much debate on the radiocarbon results for what remains of the town. In contrast with expectation based on the modern plant data, the radiocarbon results for the eruption that destroyed Akrotiri are surprisingly close to dates given by other techniques (fig. 9).

Contamination

One of the fundamental assumptions of radiocarbon dating is that no process other than radioactive decay has altered the level of ^{14}C in a sample since its removal from the biosphere. Any addition of a carbon-containing material is contamination, and it must be removed before the dating process begins otherwise a false result will be obtained. For example, calcium carbonate such as limestone can dissolve in ground water and then be deposited within a sample. Limestone, being of geological origin, has an age greatly in excess of any archaeological samples. Similarly, humic acids from burial soil can contaminate a sample, but whether the apparent result is too young or too old depends on the origin of the humic acids. Thus once a sample is accepted for dating, the first task in the laboratory is pretreatment, that is, removal of any likely sources of contamination. Pretreatment procedures are designed not only to take into account the likely type of contaminant, but also the structure of the sample. For many samples, contaminants such as carbonates are removed by an acid wash and humic acids by using dilute alkali and acid washes in sequence.

The effect that contamination has on the radiocarbon result depends on the amount of contamination present and the relative ages of sample and contaminant. The relationship between the measured ^{14}C activity (A_m) and those of contaminant (A_x) and true sample (A_s) is quite straightforward:

$$A_m = fA_x + (1 - f)A_s$$

where f is the fraction of contamination in the measured material. To convert these activities or concentrations to ages for the two constituents, however, requires substitution for A_x and A_s using the age equation:

$$A = A_0 e^{-t/8033}$$

where A_0 is modern activity (see p. 11). Hence the relation between the constituent and measured ages is not so simple. Some idea of the magnitude of the problem can be gained by considering the effect of infinitely old (i.e. 'dead') and of modern contamination. The infinitely old contamination acts as if part of the sample were missing (since $A_x = 0$); hence, for each percent of contaminant approximately 80 years' age discrepancy is introduced, the apparent radiocarbon age of the sample being older than the true one.

The effect of modern contamination is more complex to evaluate. It is usual to determine the discrepancy introduced by different percentages of modern as defined by the activity of zero radiocarbon age material ($A_x = A_0$). It must be remembered that this is the theoretical activity of a wood sample growing in AD 1950 in the absence of the fossil-fuel effect, but more importantly since the mid 1950s the level of activity of growing organic material has been significantly *greater* than this theoretical modern level due to bomb carbon (see fig. 6). The present-day level (at the time of writing) of atmospheric ^{14}C concentration is about 20% higher than it would have been in the absence of the interfering effects of humans. Hence, given actual present-day ^{14}C levels, even less contamination is required than indicated in table 2.

Table 2 *Effect of contamination (in years)*

Sample age	Contamination					
	0.1% modern	0.1% infinite	1% modern	1% infinite	10% modern	10% infinite
2500	−3	+8	−30	+80	−290	+850
5000	−7	+8	−70	+80	−670	+850
10 000	−20	+8	−200	+80	−1770	+850
15 000	−45	+8	−430	+80	−3510	+850
20 000	−90	+8	−840	+80	−5980	+850
30 000	−320	+8	−2750	+80	−13070	+850
40 000	−1080	+8	−7180	+80	−21990	+850

Example: 10% of infinite-age contamination makes a sample 5000 years old appear to be 5850 years old.

Given the errors that can arise, clearly the wisest course of action is to avoid any additional contamination in the collection, field conservation and packing of samples, particularly since some contaminants are actually impossible to remove (for example, animal glue if used on bone is chemically identical to the sample). Biocides, conservation chemicals (such as polyvinyl acetate and polyethylene glycol), cigarette ash, paper labels and wrapping paper are all sources of carbon and hence are potential contaminants.

Pretreatment

Some of the more commonly dated materials can now be considered, together with the ways in which their structure influences the way they are pretreated. The pretreatment procedure is not a fixed recipe; it is adapted as necessary to the environmental conditions and preservation of the sample.

Samples are always visually examined and rootlets are removed before any chemical pretreatment. In some samples this can be difficult since the rootlets may not be readily distinguishable from the sample. This is particularly true for peat samples that have been dried.

Wood and wood charcoal
Wood is chemically quite complex, being composed of cellulose, other carbohydrates and lignins. Of these, cellulose is the least likely to take up contamination. Pretreatment of wood for radiocarbon dating thus should ideally extract the cellulose. This is a time-consuming and potentially hazardous process involving oxidation of the structure using sodium hypochlorite. It also requires a larger sample since substantial carbon-containing portions of the wood are eliminated. It may therefore not be feasible in all situations and dates on whole wood are often produced.

By contrast, after charring or burning, the resulting charcoal (being typically about 50% elemental carbon) is relatively inert chemically though it is highly absorptive. Contamination that has entered after deposition will therefore not have chemically combined with the structure, but be interstitial and removable by acid (for carbonates) and alkali (for humic acids). The pretreatment is thus relatively straightforward.

Bone

Bone comprises, in simple terms, two fractions: a protein fraction, which provides strength and some degree of flexibility, and an inorganic component, calcium hydroxyapatite, which gives bone its rigidity and solid structure. Both contain carbon and in theory both individually (or together) are datable. The hydroxyapatite, however, is an open lattice structure into which carbonates from ground water can be deposited. Unfortunately the hydroxyapatite is also acid soluble. On the other hand, the protein component is relatively acid insoluble and can therefore be separated from the hydroxyapatite as well as from any secondary carbonate. A well-preserved bone left in dilute hydrochloric acid for a few days will leave behind a replica of itself without the rigid carbonate structure. This 'pseudomorph' is the protein fraction, comprising various amino acids, and is loosely referred to as collagen.

The protein fraction is not always well preserved, however. In particular, it begins to degrade in warm conditions and can be attacked by fungi or bacteria. When dating whole collagen it is not possible to detect these effects, but the amino acid profile (i.e. the ratios of the constituent amino acids relative to each other) can indicate if anything is amiss. With accelerator mass spectrometry (AMS) dating, since only a small sample is required, it is possible to date individual amino acids after separation to check that several give the same date. It was thought that dating one particular amino acid, called hydroxyproline, would necessarily give the correct radiocarbon age of the bone, since it was believed that this particular amino acid was specific to bone. It is now known that hydroyproline can also be a component of some ground waters and therefore potentially also present as a contaminant; hence dating of several amino acids is recommended. However, separation and dating of even single ones is time-consuming and costly and only undertaken in special circumstances or for older samples, where the presence of even small levels of more recent contamination produce a large error.

By contrast with wood and wood charcoal, the dating of burnt bone is not necessarily more straightforward than the dating of bone. In fact, only in unusual circumstances is burnt bone datable at all by radiocarbon. Collagen degrades on heating and in most circumstances of burning of bone, whether accidental in cooking or deliberate in a cremation pyre, the protein fraction of the bone is lost. On acid treatment, it is quite likely that virtually the whole of a burnt bone will dissolve, leaving a solution that cannot readily be considered free of carbonate contamination. Exceptions to this can arise if a bone was heated under reducing conditions causing carbonisation; then soft tissues such as flesh may also be preserved and be datable after pretreatment, much the same as for charcoal. Bone that has been well burnt in oxidising conditions has a fairly characteristic appearance. It can be almost white, more usually light grey, with cracking; it is also quite light in weight. Less well-burnt bone may only have a pinkish-brown colouration.

Unfortunately, the degradation of collagen subjected to heat does not require the high temperatures encountered in a fire; given time, it accounts also for the poor preservation of collagen in bone from hot, arid areas, and the combination of heat and water percolation removes the collagen from stewed bones: the heat breaks down the amino acid chains and these protein constituents can then be leached out.

Peat

Peat can be separated into three components: humins, fulvic acids and humic acids. The last two are alkali soluble, but separable on the basis of pH (the degree of alkalinity or acidity). The humins are the more solid residues of the plant remains that formed the peat, whereas the fulvic and humic acids are potentially mobile as they are soluble, depending on the acidity of the peat. These mobile components do not necessarily provide a reliable radiocarbon age for a particular peat layer, whereas the humins usually can, provided there has been no natural upheaval and inversion and provided rootlet penetration subsequent to the formation of the peat is not a problem.

Pretreatment of peat therefore requires carefully acid and alkali washes of known pH if all fractions are to be retained. It is important to note that the water content of peat can be very high and sample weights needed are consequently variable.

Shell

Mollusc shell comprises the largest proportion of shell material dated by radiocarbon. Like bone it has an inorganic (calcium carbonate) fraction and an organic fraction known as conchiolin. The latter makes up only a few percent of the total, however, and so the majority of measurements use the inorganic part. This causes problems, because carbonates are quite soluble and can recrystallise and isotopically or chemically exchange with their environment. It might be possible to circumvent surface exchanges by mechanical removal or acid dissolution of the outer layers of the shell. In some environments it may be necessary to remove as much as 50%. Radiocarbon results for the layers removed compared with that for the interior of the shell may indicate whether secondary carbonate was present.

Recrystallisation is even more of a problem, since it is not confined to the outer layers. It can be detected for some mollusc shells since, although the original form of calcium carbonate laid down would have been aragonite, on recrystallisation, calcite is formed and the different crystal structures of these two forms of calcium carbonate are distinguishable by X-ray diffraction. This is not a universal rule, however: some mollusc shells are mixtures of aragonite and calcite and others, such as oyster shell, are calcite to begin with.

As already observed, radiocarbon dating of shell can also involve other problems, such as hard-water effects and marine offsets.

— 3 —

Measurement of Radiocarbon

There are two methods of detecting ^{14}C: by conventional radiocarbon dating, which detects one of the ^{14}C decay products, or by accelerator mass spectrometry (AMS) which directly measures the number, or a proportion of the number, of ^{14}C atoms relative to ^{13}C or ^{12}C atoms in the sample. As well as archaeological samples, both also measure modern standards such as oxalic acid and 'dead' (or background) samples as reference materials. Although the principles of conventional and AMS dating are fundamentally different, both produce radiocarbon results that can be interpreted in the same way.

Conventional radiocarbon dating

Conventional radiocarbon dating techniques must not be confused with the concept of a *conventional* radiocarbon *result*, which is the method of quoting a result rather than the method by which the result is derived. An unambiguous description for these techniques would be *radiometric*; however, this term is not in common use.

The nucleus of a ^{14}C atom is unstable, so there is a finite probability at any instant in time that it will decay. When it does so, it decays to nitrogen (^{14}N) with a beta particle being emitted. A beta particle is the name given to an electron resulting from the radioactive decay of a nucleus. The beta particle can be detected fairly easily because it is electrically charged. The first detection systems devised by Libby and his group for archaeological samples used solid carbon. The sample was converted to 'lamp black' and this was coated on to the inside surface of a metal cylinder which was then inserted into a Geiger counter of a type previously designed by Libby. The sample had to be internal to the counter because the beta particles emitted by ^{14}C are of low energy and would not have been able to penetrate the wall of the counter.

Libby's experience in this field was essential to the success of the first attempts to detect ^{14}C in archaeological samples. He and his co-workers were well aware of the need to reduce or counteract the effects of any radiation other than ^{14}C in the sample. There are several sources of radiation in the environment, in particular minute but detectable amounts of thorium, uranium and potassium (^{40}K) naturally occurring in building materials, as well as cosmic rays. These are referred to as background radiation. A very significant reduction in the count-rate arising from background radiation can be achieved by shielding the counter using steel or lead with a low radioactivity content. The more penetrating cosmic rays need massive physical shielding and some counting laboratories, such as that of the British Museum, are sited deep underground. More usually, cosmic rays are dealt with by anticoincidence counting. This involves a ring of counters surrounding the one containing the sample. Any count registered simultaneously by the sample counter and the anticoincidence shield is rejected electronically, since it can only be a result of radiation that has penetrated several thicknesses of counter wall and is hence external to the sample.

Gas counting
By the mid 1950s, contamination from nuclear fall-out had become a major problem for the solid-carbon technique. There had been major advances in gas proportional counting, and since conversion of the sample to a gas also avoids the fall-out problem, the solid-carbon technique was superseded.

Although methane, ethylene and even ethane can be used, many gas-counting systems today use carbon dioxide, since this is the main combustion product of organic materials and therefore readily prepared. Care must be taken, however, to remove impurities such as air, halogens and sulphur dioxide which will affect the counting properties of the gas. Like the solid-carbon technique, lead or steel shielding as well as an anticoincidence guard are used to reduce the background count-rate. The counter itself also gives discrimination between different types of radiation, so that some types of non-carbon contamination can be electronically eliminated.

Liquid scintillation counting
During the 1960s liquid scintillation counting (LSC) became popular. The perceived advantages were that since the sample is in liquid form, it would be easier to manipulate and also the counting volume is small relative to gas counting and thus background count-rates, being volume dependent, are likely to be lower. In addition samples can be cycled in a conveyor-belt system, being counted for a few tens of minutes then removed, to return for counting some hours later. This reduces the influence of any variable effects such as the background from cosmic rays.

In LSC a scintillator is added to the sample liquid and this produces a flash of light when it interacts with a beta particle. Each flash of light is detected by two photomultipliers, devices that transform light into electrical pulses using the photoelectric effect, set on opposite sides of the vial containing the sample and scintillator solution. Two photomultipliers are used so that *coincidence* counting can be employed. Only if both simultaneously register a flash of light is this taken as a true count: radiation external to the sample will cause pulses in one photomutiplier but not the other and consequently these are largely eliminated. However, background radiation can cause simultaneous pulses, for example by interacting with the scintillator. Lead shielding around the counting chamber is therefore still needed and some counters also have an anticoincidence guard.

In the developmental stages of liquid scintillation counting, various liquid forms of the sample were tried; today only benzene is used (fig. 10).

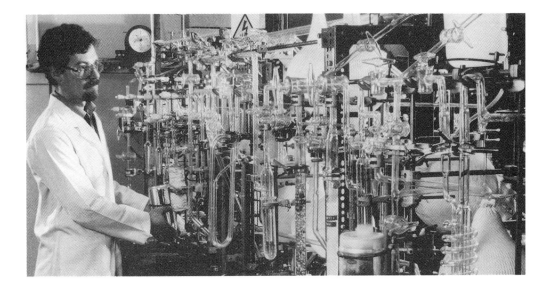

10 The radiocarbon dating laboratory at the British Museum was one of the first to be established following Libby's trial of the technique. It started in the early 1950s with gas counting, then moved to liquid scintillation counting in the early 1970s. Part of the Museum's high-vacuum system for synthesis of benzene is shown here. The sample is combusted to carbon dioxide (CO_2) which is reacted with molten lithium metal to form lithium carbide (Li_2C_2). Acetylene (C_2H_2) is then formed by hydrolysis and this is trimerised to benzene (C_6H_6) catalytically. With care, high-purity benzene can be consistently synthesised.

The growth in popularity of LSC is perhaps due most to the commercial availability of LS spectrometers, which in recent years have been designed by manufacturers with the problems of low-level ^{14}C activity in mind. The choice of gas- or liquid-counting system is, however, largely a matter of individual laboratory choice, based on available 'in house' expertise.

Small-sample systems
Typical sample sizes employed by normal gas and liquid scintillation counting are approximately the same, that is, equivalent to about 5–10 g of *carbon* (the equivalent *sample* size is indicated in table 3). Over the past decade, mini-gas counters have been in use at the Brookhaven and Harwell laboratories. These were developed to handle the gas derived from small samples yielding only about 100 mg of carbon. To achieve the same order of error term, this reduction in sample size means an approximately equivalent multiplication of counting time from about one or two days to about one or two months. To achieve a reasonable throughput of samples, about ten are counted at the same time within a single anticoincidence shield.

More recently, with the development of low-background liquid scintillation counters, two laboratories (at the Australian National University and Southern Methodist University in Texas) have tried counting small samples in specially designed small-volume vials that hold about 0.3 ml of benzene, equivalent to about 0.25 g of carbon. Smaller samples can be measured if the carbon dioxide gas they produce is diluted with a known amount of 'dead' carbon dioxide before the synthesis of benzene. This work is at an early stage; one of the

33

Table 3 *A comparative guide to sample sizes required for various ^{14}C dating methods*

Material	Conventional (g)	Mini-counting (g)	AMS (mg)
Wood (whole)	10–25	0.1–0.5	50–100
(cellulose)	50–100	0.5–1.0	200–500
Charcoal (& other charred materials)	10–20	0.1–0.5	10–100
Peat	50–100	0.5–1.0	100–200
Textiles	20–50	0.05–0.10	20–50
Bone	100–400	2.0–5.0	500–1000
Shell	50–100	0.5–1.0	50–100
Sediment, soils	100–500	2.0–10.0	500–25 000

Any guide to sample sizes requires many caveats. The amount needed depends on a number of interlinking variables such as the age of the sample, its preservation, likely sources of contamination and the precision required. The figures for bone, for example, would typically be for a 10% collagen content, normal precision, a sample age less than about 10 000 years old, and no unusual contaminants.

It is probably true to say that the lower down the list of sample types, the more difficult it becomes to give a general guide.

The guide is for dry equivalent weights, but waterlogged wood and peat should be kept wet and correspondingly higher sample sizes are needed. Remember, also, that these are the weights of actual sample, not sample plus soil and stones, and that what seems to be charcoal in the field may only be blackened soil; if in doubt look for structure in the sample.

main difficulties lies in manipulating small volumes of liquid relative to the equivalent volume, in terms of carbon content, of gas. Also, counting times cannot be greatly extended, since only one vial can be counted at a time and practical difficulties of maintaining constant counting conditions can arise due to the volatility of benzene.

Accelerator mass spectrometry

All of the techniques outlined above detect the beta particle from the decay of ^{14}C atoms. Even though ^{14}C is only one part in 10^{12} relative to ^{12}C, nevertheless 1 g of modern carbon contains about 5×10^{10} atoms of ^{14}C. Only 1% of these decay in an 80-year period, and hence in a single day of measurement there are only of the order of 10^4 disintegrations available for detection. Even though this is more than a million times smaller than the actual number of ^{14}C atoms, it is still sufficient to give a precision in the region of $\pm 1\%$ or ± 80 years (see below for a discussion of error terms).

A more efficient method for detecting ^{14}C would be to measure the number of atoms present, or a proportion of them. Direct counting of 10^4 atoms of ^{14}C to give 1% precision requires detection of only one atom in 5×10^6 of those present in the 1 g sample. Alternatively, if a milligram rather than gram sample and two hours rather than one day detection time are used, then one atom in 200 must be detected. Of course, the older the sample the smaller will be the number of atoms available for detection. Nevertheless, this still means that the detection system devised can afford to be fairly inefficient while employing a dramatically reduced sample size and counting time.

The technique by which atoms of specific elements are detected according to their atomic weights is known as mass spectrometry. However, normal mass spectrometers do not have the sensitivity to detect ^{14}C and reject all other elements or molecules of very nearly the same weight such as ^{14}N, the most common isotope of nitrogen which, comprising some 80% of the atmosphere, is very abundant relative to ^{14}C. The techniques of nuclear physics were brought to bear on this problem, and in the late 1970s two laboratories (Simon Fraser in conjunction with McMaster in Canada and Rochester in the USA) showed that ^{14}C could be detected using what is now referred to as accelerator mass spectrometry (AMS). When a magnetic field is applied to a moving charged particle, the particle is deflected from the straight path along which it was travelling. If charged particles of different mass, but the same velocity, are subject to the same magnetic field, the heavier particles are deflected the least. This is the principle of the mass spectrometer, and detectors at different angles of deflection then receive particles of different mass. The accelerator mass spectrometer works on the same basic principle, but the charged particles are

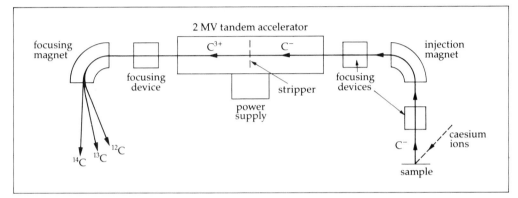

11 (*above*) Schematic diagram of a tandem electrostatic accelerator system for detection of ^{14}C atoms.
(*right*) Part of the accelerator mass spectrometry (AMS) facility at Oxford University: the ion source is in the foreground, and in the background to the left is the initial stage of the accelerator itself.

The sample is normally in solid graphite form, though carbon dioxide sources may be used. One way the graphite is formed is by combusting the sample to form carbon dioxide, which is converted to carbon monoxide in the presence of zinc and then reduced to carbon (graphite) by a catalytic reaction using iron. The graphite (typically weighing a few milligrams) is pressed on to a metal disc that, together with other sample discs, is mounted on a target wheel, enabling samples and reference materials to be measured in sequence. The wheel is referred to as a target because ions from a caesium gun are fired at it. The negatively ionised carbon atoms (C^-) produced are then accelerated to the positive terminal by a voltage difference of 2 million volts (2 MV). N^- ions are unstable and therefore cannot reach the detector. When the C^- ions encounter the stripper, electrons are lost and they emerge with a triple positive charge (C^{3+}): molecules are thus eliminated, since none can exist in this charge state. After further acceleration, this time away from the positive terminal, selection according to mass by deflection in a magnetic field takes the ^{14}C ions to a detector. ^{12}C and ^{13}C are also collected to provide the concentration ratio and allow evaluation of the level of fractionation.

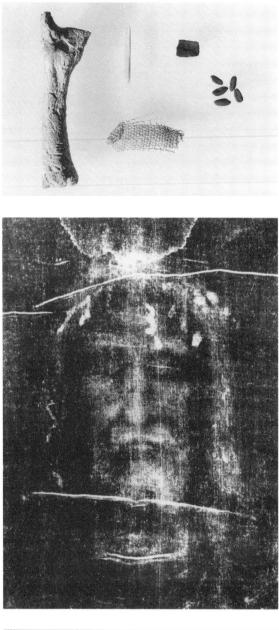

12 Mini-systems and AMS are capable of dating very small samples. The section of matchstick provides a scale (length 2 cm, weight 60 mg) as does the human metacarpal (weight 5 g, only part of which would be required). The other samples are linen (50 mg), charcoal (40 mg) and charred seeds (total weight 40 mg). AMS can date single seeds in favourable circumstances. In contrast, for dating bone, conventional methods typically require a sample weight equivalent to that of a human thighbone.

13 The Turin Shroud is a linen cloth over 4 m in length, bearing the shadowy image of the front and back of a man who appears to have been scourged and crucified; it is therefore believed to have been Christ's burial shroud. The history of this cloth is known with certainty back to about AD 1350, when it was in the possession of the de Charny family in France. Even then it appears to have caused something of a religious furore, being declared by some to be a fake and by others to be the true Shroud. In 1898, photography of the Shroud revealed that when seen in negative the image is strikingly life-like. This discovery and subsequent medical findings fuelled wide debate that the cloth could conceivably be genuine. In 1987, the British Museum was asked to participate in the certification of sampling for radiocarbon dating and in the statistical analysis of the results. Samples measuring only a few square centimetres were given to each of three accelerator laboratories: Oxford, Zurich and Tucson. The calibrated radiocarbon result, published in the journal *Nature* in 1989, was 1260–1390 AD, demonstrating that the linen of the Shroud of Turin is medieval. The result corresponds well with its first appearance in France. However, it remains to be established how this striking image came into being.

14 Mobiliary art is not uncommon from the Upper Palaeolithic in France; however, there are few examples in Britain. The horse mandible from Kendrick's Cave (Great Orme's Head, Llandudno, Wales) is one of these few. Prior to small-sample dating, objects such as this could not be directly dated unless they were totally destroyed in the process. This mandible was shown to be late Upper Palaeolithic or very early Mesolithic.

subjected to large voltage differences so that they travel at very high speeds. This enables various devices to be used to discriminate against the much more abundant elements, such as ^{14}N, and molecules, such as ^{13}CH, which would otherwise swamp the ^{14}C signal.

Cyclotrons have been used for ^{14}C detection, but the more commonly used accelerator system is a tandem accelerator, so called because there are two stages of acceleration: one towards the central positive terminal and the other away from it. This is achieved because the carbon ions are made to change polarity, from negative to positive, by a 'stripper' of gas or metal foil at this central point. A schematic diagram of a tandem electrostatic accelerator system is shown in figure 11.

The disadvantage of AMS is the high cost of establishing such a facility (about £1 million, over $1.5 million) and of running it. The great advantage of AMS over conventional techniques is of course the small sample size needed: typically a factor of 1000 smaller (see table 3). This means small objects that would be totally destroyed if dated by conventional ^{14}C can be sampled for AMS (fig. 12). Equally, very small areas of valuable artefacts or art objects can be sampled, thus minimising destruction (figs. 13, 14). A new dimension has also been added for archaeological samples, enabling dating of samples such as individual seeds that are important in the consideration of the origins of agriculture and the domestication of cereals. It has even been possible to date blood residues on stone implements. Different chemical fractions of a sample can also be dated, and this has important implications for detecting the effects of degradation and contamination on ^{14}C results. AMS also has the potential for high throughput of, say, 1000 samples per year, since each sample requires a run-time of a few hours rather than one or two days. However, sample pre-treatment has to be particularly rigorous to avoid even small levels of contamination leading to substantial errors.

Age limits

Maximum age
In conventional radiocarbon dating, the maximum age is determined by the level of background count-rate. For old samples the ^{14}C content is small and a criterion must be set for deciding whether or not the sample ^{14}C count-rate is distinguishable above the background level. The normal condition set for quoting a finite age is that the net count-rate for a sample should be more than 2σ greater than zero, where σ is the counting error (see p. 38) which, since background has been subtracted, incorporates the errors on the sample count-rate and on the background count-rate. If the net sample count-rate is within 2σ of zero, but still positive, a minimum age is often quoted; this is calculated using the net activity plus 2σ in the age equation. If the net sample count-rate is indistinguishable from zero, the result is quoted either as 'infinite' or as 'background'.

Since the upper age limit for conventional radiocarbon dating depends on the background count rate, it varies from laboratory to laboratory but is typically in the region of 40 000 years. For AMS, the upper age limit is determined by other factors, such as machine stability and the degree of modern contamination introduced in the processing of small samples. Values similar to those for conventional radiocarbon laboratories are being achieved.

Lower limit

Due to the mutual interference of the fossil fuel and bomb effects, radiocarbon results of less than 200 years are often reported as 'modern'. Of course, due to bomb-produced ^{14}C, it is possible for samples to have a ^{14}C activity substantially greater than the modern activity for AD 1950 as defined by the oxalic acid standard. Such samples may be referred to as 'greater than modern' (>modern).

Enrichment

One way in which greater ages can be measured is by using isotopic enrichment, a method which takes advantage of isotopic fractionation. There are two ways of doing this: by using a thermal diffusion column or via photodissociation using a laser beam. With the former, finite ages of about 75 000 years have been reported, but the technique is not routinely applicable since it requires a large sample (roughly an order of magnitude greater than for conventional radiocarbon) and the enrichment process typically takes about a month. The photodissociation method has been shown to be feasible but has not been employed to produce dates. It is faster but processes a much smaller sample. Its use would therefore be possible only in conjunction with AMS.

The error term

Experimental error, inherent in any experimental process, is usually evaluated by replication of the measurement process. In radiocarbon dating, time, cost and (for conventional radiocarbon) the sample size mean this is not a practical proposition. The error term is therefore estimated and then usually treated as if it were known.

When a measurement process can be repeated, the distribution of the results is usually described by the Gaussian, or normal, probability function. Assuming this holds for radiocarbon results when the error is estimated, the one sigma error term ($\pm 1\sigma$) means there is a 68.3% chance that the true result will lie within $\pm 1\sigma$ of the experimental result, a 95.4% chance within $\pm 2\sigma$ and 99.7% within $\pm 3\sigma$. The alternative view is that there is nearly a one in three chance that the true result does *not* lie within $\pm 1\sigma$ of the experimental one. Even at $\pm 2\sigma$, there is still a one in twenty chance that the true result lies outside this range. Despite this, there are many instances where radiocarbon results have been used without their associated errors, as if they were absolutely known with no uncertainty!

Unfortunately there is no convention defining how a laboratory should perform the error estimation. All laboratories include an error contribution from 'counting' or Poisson statistics. The decay of ^{14}C follows the radioactive decay law so that half the atoms decay in 5730 years (the true half-life). Over a short time interval, however, the number that will actually decay is not exactly predictable. In Poisson statistics, which are fundamental to evaluating the error in any radiocarbon measurement, the expected number of events in a given time is estimated by the measured number, n, and the standard deviation (variability) is $\pm\sqrt{n}$.

Given that the standard deviation is $\sqrt{n}$, this means counting 100 counts to achieve a 10% error in the estimate of the true number; 10 000 for 1%, and

1 000 000 for 0.1%. The longer one counts, the smaller, in percentage terms, is the error term. Liquid scintillation counting, for example, gives about eight counts per minute per gram of benzene for a modern sample. For a 5 g sample, 1% counting statistics (equivalent to an error term of about 80 years) requires counting for 250 minutes. The older the sample, the longer it takes; a half-life sample would require twice as long. Increasing either the size of the sample or the counting time will decrease the error term, but to halve the error requires a fourfold increase because of the square law relationship between error and number of counts.

In evaluating the *total* error in a radiocarbon result, however, three sets of counting statistics need to be incorporated: for sample, background and modern. Laboratories tend to differ as to which additional errors are incorporated for any non-Poisson variation. In fact, any variable in the age equation, such as fractionation correction, has an associated measurement uncertainty which should be taken into account in evaluating the overall error. The total error is found by 'propagation of errors' which 'weights' errors according to how the variables appear in the age equation. A simple example of the propagation of errors is the error on a sum. This is given by the square-root of the sum of the squares of the individual errors on the variables being summed; the errors are then said to be summed 'in quadrature'.

Strictly speaking, the error term estimated is on the ^{14}C concentration of the sample relative to that in a modern reference sample (A/A_0 on p. 11). Since the radiocarbon 'age' is proportional to the natural logarithm of this ratio, the error on a radiocarbon result is not symmetrical. The asymmetry is small except for old samples when $+\sigma$ is larger than $-\sigma$; this can be appreciated by consideration of the exponential decay curve in figure 1.

Accuracy and precision

Accuracy and precision, though often loosely used synonymously with error, are very different from each other. So far, only random errors have been considered; these determine the precision of a measurement. The accuracy of a

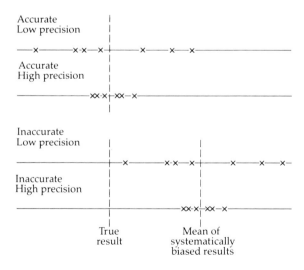

Accurate
Low precision

Accurate
High precision

Inaccurate
Low precision

Inaccurate
High precision

True
result

Mean of
systematically
biased results

15 Random errors determine the precision of a measurement; systematic errors determine accuracy. High precision means that the overall random error is small and that replicate measurements will provide closely similar results. These may not, however, be close to the *true* value they are attempting to estimate, in which case there is a systematic error giving results that are not accurate. In other words, precise results are not necessarily accurate and vice versa; this is illustrated schematically. A random error is one which, when averaged, tends to zero as the number of measurements averaged is increased. Hence, by combining the results of replicate measurements, in theory a more precise mean result can be obtained. A systematic error persists through a series of replicate measurements and cannot therefore be reduced by averaging; it produces a bias relative to the true result.

39

measurement is determined by systematic errors. Figure 15 schematically illustrates the difference and shows that, without accuracy, high precision is valueless.

High-precision dating

Given the number of factors that need to be taken into account in evaluating a radiocarbon result, the care taken in achieving high-precision results with error terms less than ± 20 years can begin to be appreciated.

Higher than normal precision can be achieved by increasing sample size and counting time, though there are practical limits on both of these. In principle it is also possible for a result with higher precision to be achieved by replicate measurements of a sample. If a laboratory with a normal precision of ± 50 years were to try to improve its precision to ± 25 years, in theory it would need to do four replicate measurements. The alternative would be to submit the whole, large sample to a high-precision laboratory that would provide in one measurement a result with a precision of ± 20 years or rather better. There is one obvious advantage to this course of action: one high-precision result does not cost as much as four normal-precision results. It is also probably fair to say that most normal-precision laboratories do not usually evaluate all potential sources of random error but only the major ones. The contribution from unaccounted errors may be small relative to the typical error from such a laboratory, but it will increase as attempts are made to achieve more precise results. High-precision dating is therefore not simply a question of increased sample size, more counting time, or replication: as well as minimising and evaluating *all* sources of random error, high accuracy must be achieved and maintained.

Not all archaeological samples warrant high-precision dating, however. In many situations the association of the sample with the event to be dated may be poor, or there may be an unknown age offset.

Inter-laboratory comparability

Laboratories test whether or not they have systematic errors by taking part in intercomparisons. Several samples of different age and material are usually included in the test, and the laboratories date portions of each. The majority of laboratories should be in good agreement; however, there is often surprising variability. For each sample in the test, a consensus of the results of the participating laboratories is taken as the true value. Those laboratories which are offset by amounts significantly greater than their random error estimate may simply be underestimating this error, but more probably they have systematic errors. Systematic errors can occur in many different ways depending on the method used to measure the ^{14}C content. In liquid scintillation counting, for example, care has to be taken to avoid evaporation of the benzene. An unsuspected loss of 1% of the modern reference will give sample ages too young by about 80 years. Continued evaporation will give even larger biases, but the individual samples dated while this is happening will not necessarily be in error by the same amount; it all depends on when each one was measured relative to the evaporating reference sample. Hence errors can be systematic *and* variable in magnitude.

The majority of laboratories conduct continual self-checks that should indicate whether systematic errors are likely to be present, so that any drift can be rectified immediately. The purpose of intercomparisons is to give laboratories the opportunity to perform independent checks.

Error multipliers

There are at least two types of error multiplier in radiocarbon dating. These are *ad hoc* ways of attempting to make estimates of error on a radiocarbon result more realistic, that is, more in line with observed variability in results. Ideally, variability, and therefore the quoted error on a radiocarbon result, should be evaluated from several replicate measurements of a sample. However, this is usually impractical, so an estimate of the error is given instead. Some laboratories repeat radiocarbon measurements on one or two samples, which they process at intervals over a long period of time. This series of measurements will test for random variables that may only become apparent on a long time-scale. If the overall variability is greater than the individual estimates of error, the laboratory will apply an error multiplier to error estimates for all samples.

The other type of error multiplier is one deriving from a laboratory's performance in an intercomparison and is the factor by which the quoted errors would need to be multiplied in order to bring them into line with the laboratory's observed variability relative to the consensus results. If the laboratory has not underestimated its random errors but has an offset due to a systematic error, then an error multiplier does not adequately express the problem. The purpose of an intercomparison is largely to enable laboratories to test whether they have significant systematic errors, and if so to identify the problem and rectify it. Hence this type of error multiplier, as well as being potentially misleading, should not be needed since the situation will not be static.

It is important that users of radiocarbon results be aware of the limitations and possible pitfalls of quoted error terms. But many laboratories are scrupulously careful in their attempts to evaluate all sources of random error and to avoid systematic errors by continual self-checks and participation in intercomparison studies. If there is any doubt, a laboratory should be able to provide data to demonstrate its reliability.

Errors due to time-widths and age offsets of samples

Before leaving this subject, there is one further aspect of errors that needs to be raised. There is a mistaken belief that the error term on a radiocarbon result takes into account the errors introduced by the inherent time-width of a sample and by age offsets resulting from a difference between the time of death and time of ceasing to exchange with the biosphere. It does not. It only evaluates the random error on the measurement of the overall ^{14}C concentration, whereas these effects largely introduce systematic errors and in many cases are not quantifiable. (The effects of such offsets and time-widths are considered qualitatively in chapter 5.) In addition, it is important to note that radiocarbon dating of an artefact is not necessarily the same as dating an archaeological event (this too is discussed in chapter 5).

If there is a marine reservoir correction to be made, this should be added to the radiocarbon result and an overall error term calculated, taking into account the uncertainty of the correction.

Citing radiocarbon results

There is a recommended convention for citation of radiocarbon results, the elements of which can be outlined as follows:

- Radiocarbon results are given in uncalibrated years BP, where 0 BP is defined as AD 1950.
- The half-life used to calculate a radiocarbon result is the Libby half-life of 5568 years, not the more accurate value of 5730 years.
- Results are calculated after normalisation of $\delta^{13}C$ values to $-25‰$.
- Modern activity is defined as a set proportion of the activity of one of the US National Bureau of Standards oxalic acid standards (74.59% of the new standard when its $\delta^{13}C$ is normalised to $-25‰$).
- An error term of $\pm 1\sigma$ should be quoted.
- Rounding of results and error terms is to the nearest ten radiocarbon years for error terms greater than ± 50 years, and to the nearest five if smaller.

Each result also has a laboratory reference number that should be given whenever the result is quoted. Such reference numbers have a laboratory identifier (e.g. BM for the British Museum) followed by a hyphen and then a number. For instance BM-2558 is the unique identifier for a British Museum radiocarbon result on wood from one of the rungs of a rope ladder found in the tomb of Sethos I (Valley of the Kings, Thebes, Egypt); the result was 2020 ± 50 BP. Where a date is published in the journal *Radiocarbon*, the entry reference should also be quoted (for the example given: 1989 *R*, vol. 31, p. 23), then the sample details and a comment on the result can be located.

BP is variously referred to as 'before present' or 'before physics', but both mean AD 1950. One point to be aware of is that other dating techniques may use a BP notation that is differently defined, as for example AD 1980 used in thermoluminescence (TL) dating.

It may seem odd to use the wrong half-life to calculate radiocarbon dates. However, provided the convention is adhered to, no misunderstanding should arise. It has the effect of producing radiocarbon results that are 3% too young; this is automatically adjusted for in the calibration of a conventionally calculated date, because the radiocarbon results in the calibration curve are calculated on this half-life as well. If the result is beyond the range of calibration, this 3% needs to be added, though of course there is still the problem of an unknown offset from calendar years. The Libby half-life was adhered to because many radiocarbon results were produced in the 1950s before the better estimate of half-life was available; for reasons of comparability, and to avoid errors arising due to half-life corrections being made more than once, it was decided to retain the old value.

The reasons for normalising to a specific $\delta^{13}C$ value, and for use of an artificial standard to define modern activity, are discussed in chapter 2.

— 4 —

Calibration of Radiocarbon Results

The need for calibration

Various factors other than radioactive decay can affect the concentration of ^{14}C in plants or animals; these were considered in chapter 2. Source effects resulting from marine origin, hard water and volcanoes may not be relevant for the majority of archaeological samples. With alteration effects, a correction can be made for fractionation, recrystallisation can often be identified and the affected samples rejected, and *in situ* production is unlikely to be significant. Nevertheless, there are some influences that are both global and pertinent to all samples and thus can neither be avoided nor circumvented by careful choice of context or sample: these are the production effects. They are not insignificant in magnitude, having at some periods in the past accounted for a discrepancy of some 900 years between radiocarbon results and true calendar years.

Fortunately, these discrepancies can be evaluated to enable radiocarbon results to be calibrated. Since production effects are rapidly distributed throughout the atmosphere, a curve of radiocarbon 'age' versus calendar age for one material and one geographical region will serve as a global calibration curve. Sample specific reservoir effects aside, the only global difference is a northern-to-southern hemisphere effect of 30 years (subtract 30 years from dates for the latter before calibration).

Arnold and Libby's 'curve of knowns' was the earliest plot of radiocarbon results versus known age. It was for a restricted age range, but more importantly it did not show the discrepancies in radiocarbon relative to calendar age, and serves to illustrate the other requirements of a calibration curve: the known ages must be completely defined without any doubt, and the radiocarbon results must be accurately and precisely determined. Historical dates may not be particularly well known, the reigns of particular kings may 'float' by several decades depending on the precise interpretation of the historical record. Even if the endpoints of a reign are reasonably well defined, an 'event' occurring within the reign may not be. By contrast, dendrochronology provides the ideal basis for the known-age axis.

Dendrochronology

The origins of dendrochronology lie in climate studies, rather than in a need for a dating method. In temperate climates, trees grow by the addition of an annual ring, but the width of each ring varies depending on climatic conditions such as temperature and rainfall (fig. 16). For a living tree, counting backwards from the cambium layer gives the age of a particular ring, and its relative thickness indicates whether in that year the growing season was good or poor in that locality. Some trees respond less than others to environmental conditions, and are referred to as complacent when their ring widths vary little. Non-complacent trees of a single species growing in the same locality should have a similar temporal pattern of ring widths which is uniquely defined, like a signature, by their common history. This is the basis of cross-dating: being able to associate, on the basis of duplication of pattern, a tree-ring sequence of unknown age with one of known age. This enables long chronologies, or 'master curves', to be established as illustrated in figure 17.

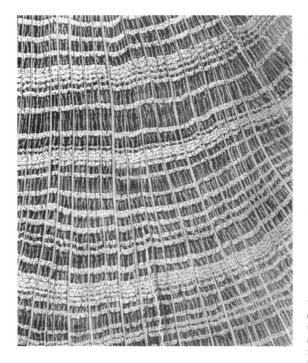

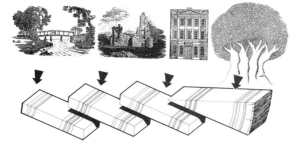

16 Trees grow by the addition of rings. The growth region is a thin band of cells called the cambium, lying between the bark and the sapwood. Division of these cells adds new bark to the outer side of the cambium and new sapwood to the inside. In a tree such as oak, the first layer or two of cells produced in spring are large vessels to transport water. During the summer the cells are smaller and the growth is more fibrous and dense. As growth finishes for the year, a layer of very small cells is produced. Therefore in cross-section the trunk of an oak has particularly well-defined rings, as this polished section shows. In temperate climates where there is a contrast between the seasons, rings are normally added annually. For many species of tree, oak included, the width of the ring varies according to the local climatic conditions. Annual rings and ring-width variation are fundamental to dendrochronology.

17 Long chronologies or master curves are established by starting with living trees or timbers where the zero-age ring is present and the year of felling known. The timescale is then extended, using large felled timbers with ring-width patterns sufficiently overlapping the existing chronology to be certain of a unique match. When no overlap is found, a floating chronology may be formed; this is a sequence, possibly built up from several timbers, the position of which in time is not known. Only if timbers providing the missing link are found can the floating chronology be tied down. Radiocarbon can provide approximate ages for timbers to show whether they are likely to be of value in linking or extending existing chronologies.

In dendrochronology the timescale is accurate to one year. To achieve this, several ring patterns are averaged for each section of the chronology to ensure no errors are introduced by the growth quirks of individual trees. Hence a great deal of hard graft, intelligent guesswork and good fortune goes into establishing a continuous chronology.

Given the accuracy of dendrochronology, it is pertinent to ask why it is not always used for dating wood. Not all tree species show climatic variability in their ring widths, and those that do respond to the local climate in which they are growing, so that a master chronology is needed for the relevant region as well as for the genus of timber. In addition, to match a ring sequence to a master chronology requires a minimum of about a hundred rings to ensure that the pattern is unique. It is relatively rare for archaeological deposits to produce wood or charcoal with this number of rings and most sites will never be datable by dendrochronology. It does find many applications in the accurate dating of buildings and phases of rebuilding, as well as in authenticity testing of objects such as panel paintings. If it can be demonstrated that the purported artist died before the year of felling of the tree used for the panel, then the painting is clearly a fake.

Calibration curves

Several long chronologies now exist for different species of tree and different localities, for example, over 8000 years for the bristlecone pine (*Pinus aristata*) in California (see ch. 2) and more than 7000 years for oak (*Quercus* sp.) in Ireland. The former was used by Suess in establishing the first useful calibration curve and the latter for the Belfast high-precision curve shown in figure 5. Once the calendar (dendro) timescale has been produced, groups of ten or twenty rings are dated by radiocarbon to provide the y-axis of the calibration curve.

Ideally, to establish the fine structure of the calibration curve, single rings should be dated with high precision, and indeed the Groningen laboratory has done high-precision measurements on single rings of German oak. However, several factors militate against establishing such a curve over a long period of time. The first is the work and time involved to undertake 8000 or more high-precision measurements. Second, with some tree species, it would be difficult to get sufficient sample from a single ring to enable high-precision measurements to be made (for oak the typical ring width is about 1 mm, but for bristlecone it is roughly a quarter of this). Third, many archaeological samples have an inherent time-width of greater than ten years (see below).

In the fifteen years following the production of the first calibration curve, a great deal of work was done attempting to establish whether or not the wiggles drawn by Suess using 'cosmic schwung' (see p. 18) were valid or a product of the imprecision of the measurements. A bewildering number of calibration curves appeared, together with an equally confusing number of statistical interpretations and compilations of the curves. These have now been superseded, in the period back to 2500 BC at least, by curves produced by Gordon Pearson and Minze Stuiver.[1,2] Their use is recommended by the international radiocarbon

1 Pearson, G. W. and Stuiver, M. 'High-precision calibration of the radiocarbon time scale, 500–2500 BC'. *Radiocarbon*, 1986, v. 28, pp. 839–62.

2 Stuiver, M. and Pearson, G. W. 'High-precision calibration of the radiocarbon time scale, AD 1950–500 BC'. *Radiocarbon*, 1986, v. 28, pp. 805–38.

community because two high-precision laboratories, Belfast and Seattle, using different conventional radiocarbon techniques and different tree species, have independently produced curves in agreement to within a few years for each sample of corresponding twenty tree rings. This is a major achievement. Other high-precision curves also exist but are yet to be verified by a second laboratory. In particular, the curve of the Belfast group extends back to 5210 BC.[3] Beyond 5210 BC, there are more curves in the same calibration volume of *Radiocarbon* which cover different periods of time, though, unlike the Belfast curve, none is continuous and based on high-precision dating of only tree rings.

Calibration of radiocarbon results

The calibration curve is not a monotonic function; that is, as true age increases, radiocarbon age does not necessarily increase. It may in fact decrease as a consequence of the wiggles (see fig. 18). If the Gaussian or normal distribution of radiocarbon result and associated error term is transformed by such a curve to the calendar axis, the distribution of calendar dates is no longer Gaussian, nor is it mathematically definable, and its form will depend on the part of the calibration curve under consideration. Calibrated dates are therefore not central dates with an error term, but a range or ranges. In fact, there is currently no consensus opinion on exactly how to calibrate a radiocarbon result, though this is under discussion. First the data points of the calibration curve must be joined up. Straight lines (see figs. 18 and 23) are usually adequate, but are not representative of natural processes: the alternative is computer-produced curves called spline functions. Then there are basically two approaches to cali-

3 Pearson, G. W. *et al*. 'High-precision ^{14}C measurements of Irish oaks to show the natural ^{14}C variations from AD 1840–5210 BC'. *Radiocarbon*, 1986, v. 28, pp. 911–34.

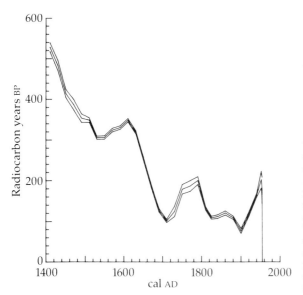

18 Section of Stuiver and Pearson's high-precision calibration curve for the recent past. Wiggles such as are shown here mean that a single radiocarbon result can correspond to more than one calendar result, as illustrated schematically in fig. 19.

19 (*opposite*) The intercept method for calibration of a radiocarbon result, $t \pm \sigma$, finds the points of intersection of $t + \sigma$ and $t - \sigma$ with the calibration curve, where σ includes the laboratory error on the radiocarbon result and the error on the calibration curve combined in quadrature. These schematic illustrations show the effect of calibration of a result when the curve has different forms. Note the occurrence of multiple ranges where the calibration curve is 'wiggly': increased size of the calibrated range relative to the uncalibrated one where the slope of the curve is effectively less than 45°, and decreased size where the slope is steep.

bration itself: one is the intercept method and the other is probabilistic. In both, the error term on the calibration curve is first added in quadrature (see p. 39) to the error term of the radiocarbon result to be calibrated.

The error term on the calibration curve at any given point in time can be read off from the curve. The values for the recommended curves are all ± 16 years or less, and some are as little as ± 3 years. Any marine correction must be added before calibration and the overall error term adjusted accordingly. For southern hemisphere dates, 30 years are subtracted before using the recommended (northern hemisphere) curves.

The intercept method

This is the method used to provide the calibrated date ranges tabulated in the papers of Pearson and Stuiver (see p. 45). Where end-points of the $\pm 1\sigma$ (where σ includes the calibration curve error) range intersect the curve, these are taken as the end-points of the 68% probability range(s) of the calibrated date. Three examples are schematically illustrated in figure 19.

It must not be assumed that the highest probability is in the centre of the range. The probability approach attempts to quantify the distribution of the calendar dates.

The probability methods

The intercept method does not fully utilise the data: it does not take into account the Gaussian distribution of the uncalibrated result, so that all dates in the calendar ranges seem equally likely. The probabilistic approach attempts to remedy this shortfall. Various methods differing in detail have been used, hence no single one has been recommended at the time of writing. The broad principles of each are the same, however, and are illustrated in figure 20.

The effect of time-width on calibration

Calibration curves are constructed using dendrochronology for the x-axis. Groups of tree rings are then radiocarbon dated and the radiocarbon result for each group is assigned to the centre point. The size of the group can be from one upwards. The practical choice is usually ten or twenty, as discussed above. The time-width of the sample used for constructing the calibration curve has implications for calibration of samples with different time-widths. If a single-year calibration curve were to be used for a sample with, say, a 25-year growth period, the curve would need to be smoothed, because the effective ^{14}C in

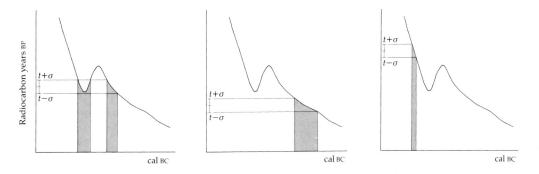

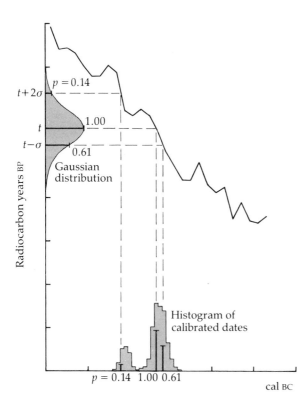

Radiocarbon years BP

$t+2\sigma$ — $p = 0.14$

t — 1.00

$t-\sigma$ — 0.61

Gaussian distribution

Histogram of calibrated dates

$p = 0.14$ 1.00 0.61

cal BC

20 The probabilistic calibration methods attempt to take into account the Gaussian distribution of the uncalibrated result. At the time of writing, no single approach has been internationally agreed, but the broad principles of each are the same. Since the calibration curve cannot be described by a mathematical formula, each approach relies on computer methods. The spread in calendar age corresponding to the error in radiocarbon measurements is simulated by calibrating the central result and point dates at intervals (say, yearly) from the centre. To each corresponding date the associated Gaussian probability (p) is attached. The calibrated dates are then grouped together, for example in ten-year segments, and the probabilities within each segment summed to form a histogram. The most probable calibrated date ranges are found by grouping segments together for the required level of certainty.

the sample is an average of the contributions of the individual years of the growth period. Similarly, decade or bi-decade curves need to be smoothed if a date on wood with, say, a hundred-year time-width is to be calibrated. Some of the curves in the calibration volume of *Radiocarbon* (1986, vol. 28) are replotted after such smoothing.

Conversely, decadal or bi-decadal curves are inappropriate for calibration of samples with a single season's growth, such as grasses, or with only a few years of growth, such as small twigs or bone from young animals. Whenever possible a single-year curve should be used, then a decadal one in preference to bi-decadal. One suggestion has been to apply an extra error term which, given the work on sunspot activity mentioned in chapter 2, has been suggested as ±15 years for single-year growth, falling to zero for an age-width of 20 years. This error is to be added in quadrature to the measurement error.

Wiggle-matching

It is occasionally possible to date accurately some materials to within about ten *calendar* years. This requires high-precision dating of several samples of tree rings from a timber so that the true temporal interval between the radiocarbon results is known. A floating piece of calibration curve is thereby produced which can be fixed in calendar time by matching it to the master calibration curve. Of course, there is still the need to ascertain the date of felling of the timber by ring counts from the cambium layer, or by estimation if the bark–sapwood boundary is missing.

Wiggle-matching might also be applicable to the dating of sediment layers if the deposition is annual.

Citing calibrated radiocarbon results

As for radiocarbon results themselves, there is also a recommended convention for citing dendrochronologically calibrated results. In calibration, radiocarbon years are converted to calendar years, but to avoid confusion it is important to distinguish calibrated dates from true historical dates. The citation convention needs to be unambiguous, hence the recommended use of cal BC, cal AD and, if necessary, cal BP: calibrated radiocarbon results must not be given simply as BC, AD or BP.

It is important to quote the uncalibrated result, the curves used for calibration, as well as the method of calibration, and to indicate any corrections that have been made to the original result before calibration. It is also important to say what confidence level corresponds to the calibrated ranges. Although it is conventional to quote raw radiocarbon results with $\pm 1\sigma$ errors, users can choose to use the 95.4% confidence level for calibrated dates if they wish. It is worth remembering that there is nearly a one in three chance of the true result lying outside the 68.3% confidence range(s), and it makes much more sense to cut the chance to one in twenty by using the 95.4% range(s). Simply double the overall error term, unless the laboratory has indicated that the errors are non-Gaussian.

Although not recommended for use by the radiocarbon community, a convention popularly used by British archaeologists in particular is mentioned here for the sake of completeness. Small letters were used to indicate *un*calibrated radiocarbon results, namely bc, ad and bp. Unfortunately, the corresponding capital letters were simply used for calibrated dates and hence could cause confusion with historical dates. In the recommended convention, the bc, ad and bp notations are not all given equivalents: only BP is used for uncalibrated results. This is rather unfortunate for many archaeologists who have long worked with radiocarbon results and conceptualise the broad currency of, say, certain building styles or object types in terms of a number of centuries bc or ad.

Perhaps with the benefit of hindsight it might have been preferable if radiocarbon measurements had never been expressed as 'ages' or 'dates'; then there could be no misunderstanding.

— 5 —

Radiocarbon and Archaeology

The archaeological record is an incomplete and fragmentary version of past human activity. What was deliberately, inadvertently or incidentally left behind is only a part of the material aspects of that activity, and this partial record has itself been subject to the vagaries of preservation and subsequent natural or human activities. The archaeologist is therefore faced with an incomplete and unrepresentative set of data from which a coherent whole must be inferred. A process of logic is used to link past events with contexts and features, such as stratigraphic levels and post holes, and to link these with artefacts found within them. If the artefact is organic it can be radiocarbon dated, but it is rare that a date for the artefact *per se* is required; instead it is assumed that the radiocarbon result will also date the event.

In many cases this may not be an unreasonable assumption. In the dating of a bone from an articulated skeleton in a grave, the assumption of association of sample and context (i.e. bone and grave) and of contemporaneity of sample and event (i.e. bone and burial) are good. All too often, however, if the samples submitted for dating are even to begin to answer the chronological questions being posed, the stages of inference linking event with context and context with artefact need more careful examination, together with the implications of what is represented by the ^{14}C activity of a sample. Liaison between archaeologists and radiocarbon scientists is therefore required from the planning stage of an excavation in discussing what radiocarbon can and cannot do, as well as practicalities such as sample size and packing. The better the liaison before and during excavation, the more likely it is that a useful series of samples will be processed.

The following sections elaborate on these points for the user, or potential user, of radiocarbon dating.

The axiomatic sample–context relationship

Deposition of any organic material in the ground obviously postdates the formation of that material and the cessation of its exchange with the biosphere. All radiocarbon age offsets make samples older than their usage or removal from the biosphere, and some, such as marine and 'old-wood' effects, make them substantially older. The exception is contamination, which can make samples appear older or younger, but pretreatment is designed to remove this. Furthermore, all depositional processes, other than downward movement as through animal burrowing or root action, are such that a date for a sample pre-dates the context in which it was found. Hence all radiocarbon samples provide a *terminus post quem* ('date after which') for their find context. How much they pre-date the deposit depends on both the nature of the sample and the taphonomic processes involved.

The 'old-wood' problem

Samples can appear to have a significant age at death due to reservoir effects such as hard water, or marine or volcanic origin of its carbon (see ch. 2). However, the more commonly encountered cause of an apparent age at death is when the organism ceased exchange with the biosphere before death, as in the case of wood (see ch. 1).

Great care must be exercised in the selection of wood or charcoal for radiocarbon dating. If the sapwood to heartwood boundary is identifiable, the age offset can be estimated using ring counts, or can be minimised by dating sapwood alone. Indeed, if sufficient rings of appropriate wood are present, dendrochronological dating may be better than radiocarbon (see ch. 4). Alternatively, twiggy material (identifiable if the complete cross-section is present by the presence of sapwood, the small number of rings and the curvature of the sample) is best since the age offset will then be small and seasoning or re-use of such material is unlikely. It is highly advisable that a specialist identify the tree species from which the wood or charcoal derived, since this will indicate whether the species was long-lived and hence whether a significant age offset is likely. If a mixture of species is represented, short-lived ones can be separated out and dated.

When there is no alternative to dating material derived from long-lived species, it is important to ask whether the result will be useful and therefore whether the sample is worth submitting. In some circumstances mature oak may be quite helpful in providing an approximate date for a monument. However, a sample of a long-lived wood species should not be considered if it overlies the context to be dated. Samples with an unknown age offset cannot provide a *terminus ante quem* ('date before which') for the deposition of the underlying context.

Quite often this 'old-wood' problem is inadequately considered by those who submit radiocarbon samples. Perhaps if bristlecone pines and yew trees, with potential longevities of about 4000 and 1000 years respectively, were to feature more in the archaeological record, the problems would be more readily appreciated!

Association

Apart from the importance of dating adequately sealed and unmixed contexts, there are also various calibres of association between the sample and the event to be dated. These were elucidated in the early 1970s by H. T. Waterbolk, a Dutch archaeologist, but his sound ideas often seem to be overlooked in the

21 In 1984, following peat-cutting operations, the upper body of a man was found at Lindow Moss (near Wilmslow in Cheshire, England). Owing to the preserving properties of the peat, a range of forensic as well as archaeological techniques could be applied, and it was discovered that Lindow Man appeared to have been ritually murdered. He had been garrotted, his throat cut and he had also received two severe blows to the head. Radiocarbon dating was applied to small samples of various types from the body itself, and two techniques, accelerator mass spectrometry (AMS) and mini-gas counting, were used. Unfortunately, although the association between the samples and the event to be dated were good, the agreement between the two techniques was not. AMS suggests Lindow Man was killed sometime in the first century AD, whereas mini-gas counting suggests that the event occurred some three or four centuries later: this is surely a mystery equal to that of the motive for the murder itself!

pursuit of dates. The best association is obviously when a date for the sample itself is required and age offsets are small. For example, in the dating of a bog body such as Lindow Man (fig. 21), a date for the body is required rather than a date for the bog in which it was found. The most dubious of associations can arise because the processes by which the sample and deposit have been brought together are ill defined or poorly understood. This is exacerbated by situations where dispersed material is bulked together to provide a 'single' sample for dating.

Mobility of samples is also a factor that needs to be considered now that facilities exist for processing very small samples. A small fragment of bone is more susceptible to movement by natural and anthropogenic mechanisms than a large bone and should not be dated in preference simply on the basis of size. If there is some reason for not destroying the intact bone, then a small sample can be taken from it for AMS or mini-counting. On the other hand, there may be good reasons for dating single grains, despite the danger of mobility, if the grain is identified to species and its presence in the context is of major agricultural significance.

The archaeologist is of course best placed to judge the reliability of association of sample and context, using the guiding principles of definable archaeological processes, selection of coherent samples rather than bulked scatters, and assessment of the likelihood of intrusive material.

Delayed use, re-use and residuality

Age offsets inherent to the sample material have already been discussed (see p. 51). Here offsets are considered that are some function of past human behaviour. The effects of these depositional processes are by no means quantifiable, but each can result in a sample giving substantially too great an age for the context being dated, even when the apparent association is good.

Delayed use

The idea of delayed use is familiar for wood, where seasoning might be involved prior to actual use of the timber. A less obvious example is the use of driftwood, particularly where indigenous building material is scarce. Here the identification by species might indicate the use of a foreign wood.

The custom of peat burning could also give large offsets, due to the use of aged material, if sediment samples from some sites were dated. The same would apply to coal, though here the radiocarbon age of the material is infinite, as it is for bitumen, a natural product of coal deposits. The use of bitumen is known at some Neolithic sites in the Near East, being used for decorative purposes as well as utilitarian ones such as the water-proofing of baskets.

Re-use

As the historic buildings of the relatively recent past demonstrate, hardwoods in particular are resilient to decay and the re-use of large timbers in rebuilding

22 Re-used timber found in association with a Bronze Age trackway at Withy Bed Copse in the Somerset Levels (England). Morticed timbers and worked wood have been found as 'make-up' beneath Bronze Age trackways of both the Somerset Levels and Ireland. Working would have had no utilitarian function in laying or stabilising the track; rather it indicates re-use of material from defunct structures. Often, however, re-used materials do not bear recognisable signs of a former use and might mistakenly be assumed to be contemporary with the context.

or other ways is not unexpected. This can be recognised if working of the wood inappropriate to its last usage is apparent (fig. 22).

Residuality

Residuality is used here to describe the incorporation of material of an earlier phase of activity in a later archaeological deposit. It therefore encompasses objects that stay in use for a considerable time or remain on the surface prior to incorporation in an archaeological context: broadly speaking, 'heirlooms' and unburied rubbish. It also includes material that has been disturbed and redeposited, but not obviously mixed layers due, say, to animal burrows. Residuality can be difficult to recognise in the archaeological record, but negative evidence can help in some circumstances. For example, articulated bones are unlikely to be residual, whereas scattered weathered deposits such as might be found in secondary ditch fills are of more dubious value as dating material.

Liaison with the radiocarbon laboratory

The fortunate few with unlimited finances will always be able to find a commercial laboratory that will date any and all samples submitted. Whether this is a wise strategy is another matter. Certainly, if funds are more modest, or non-existent, it is especially pressing that care be taken to select the most appropriate samples for answering the key questions. To do this, advice should be sought from the radiocarbon laboratory that is likely to be dating the samples. It is important that the radiocarbon laboratory be consulted *before* samples are taken, for several reasons as outlined below:

- Type of sample. Not all laboratories process all types of radiocarbon sample. For example, some do not date carbonates, others might not be willing to accept peat.
- Sample sizes. Each laboratory will advise on the amount of a given material that *it* ideally likes to process.
- Packing. Again, laboratories might have individual preferences with regard to certain types of material such as waterlogged wood; some might like to receive it after drying, others may prefer it wet.
- Systematic and random errors. The potential submitter can ask whether there is any bias in the results of the laboratory and whether the error term is a realistic estimate of the random errors.
- Error terms and calibration. The laboratory should be able to say what approximate size of random error term will be achievable for different sizes of sample. For the expected age of the sample, the likely calibrated date range(s) can then be evaluated. This information will help decide whether radiocarbon can provide the chronological resolution required, or whether higher precision results are needed, or indeed whether too much is being expected of the method.
- Cost. A laboratory will provide a costing which might vary according to the type of sample dated or precision required; even laboratories that do not normally charge might require payment if the samples do not fall within their current research interests.
- Timescale. Radiocarbon dating is not an instantaneous process; laboratories often have a waiting list.

- Identification. The dating process is destructive; the radiocarbon laboratory may be able to advise on the types of identification that should be done, but it is the responsibility of the submitter to ensure that they are complete before the dating begins.

Information is of course a two-way process, and the laboratory can learn what types of contaminant may occur on a site and hence whether special pre-treatment procedures are needed. Contaminants can be of two types: carbon-containing materials (discussed in ch. 2) that would change the apparent radio-carbon age of the sample, and other chemicals, particularly sulphur compounds, that make it difficult to process the sample and produce a pure derivative from the carbon.

A radiocarbon laboratory will also ask what is the expected age of the sample. This is not cheating! There are two reasons for asking. The primary one is to ensure no 'memory effect' in the processing of a sample: laboratories endeavour to avoid cross-contamination, but any small effect will be negligible if samples of similar age are processed in sequence; in particular, samples of substantial age (> 10 000 years) must not follow modern ones. The second reason is to avoid dating samples where radiocarbon will be of little help unless the age is a complete unknown; for example, in the period 800–400 BC, the calibration curve is effectively flat (fig. 23) and all calendar events in this period will produce approximately the same radiocarbon age.

Collecting and packaging of samples

It is important not to introduce any contamination when collecting and packing the sample. If flotation is used in the collection process, no hydrocarbons should be used. Hydrogen peroxide can, however, be used to break up soil samples.

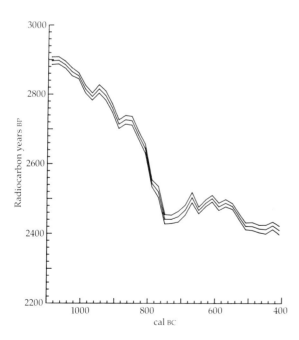

23 Pearson and Stuiver's high-precision calibration curve for 1100–400 BC shows that for the later part of this period, corresponding in Britain to the Early Iron Age (EIA), the curve is effectively flat. Calibrated dates for EIA sites therefore have calendar date ranges of about four centuries; this is of little value in British archaeology at this period since more refined typological chronologies exist. In contrast, immediately preceding 800 BC, the curve is unusually steep and therefore dating appropriate material, particularly with high precision, leads to very small calendar ranges.

Many materials used for preserving or conserving samples contain carbon that may be impossible to remove subsequently: do not use glues, biocides, polyethylene glycol (PEG) or polyvinylacetate (PVA). Many ordinary packing materials, such as paper, cardboard, cotton wool and string, contain carbon and are potential contaminants. Cigarette ash is also taboo.

The other point to ensure is that the packing and essential labelling will survive transportation and, if necessary, long storage. It is very frustrating for everyone if rubber bands have perished and allowed the sample to escape or if the sellotape that once firmly held the sample details gives up the ghost and is lost! It is also quite remarkable how rapidly some types of inks can fade or rub off the outside of polythene bags.

It is best to double or treble bag samples in strong self-seal polythene bags. Labels can then be placed between the outer skins of the packing material and, to be really cautious, the labels can also be bagged. Glass containers can be used, but they are liable to breakage. Some archaeologists use aluminium cooking foil for wrapping samples. It can be very difficult finding all the fragments of even a 10 g sample of charcoal in the many folds and crinkles, so never wrap an AMS sample like this. Also acidic samples like peat will dissolve the foil and be lost. For AMS samples, there is the possibility of contamination by plasticisers if the sample is wrapped in some plastics. Consult the laboratory or put the samples in screw-top aluminium containers.

24 Late Bronze Age waterlogged wood from a site on the Thames bank at Runnymede Bridge (Berkshire, England). This timber was part of a row of piles driven into the edge of a contemporary river to support a palisade around the settlement. This row of piles has been dated by eight radiocarbon measurements giving a mean value of 2740 ± 30 BP. These should date the construction closely since all the piles were fashioned from young trees.

Waterlogged samples are an interesting problem: should they be dried before submission or sent wet? Charcoal should be dried. Bone if not too friable should be washed and dried, but remember to do this at normal ambient temperatures: drying in an oven will degrade the collagen component without which the bone cannot be reliably dated. Both waterlogged wood and peat (fig. 24) should be submitted wet, and if there is likely to be a long time between the collection of the sample and submission for dating, it should be frozen to avoid unpleasant infestations. If peat is dried it becomes impossible to distinguish the modern rootlets from the structure of the sample. The reason for keeping a large quantity of wood wet is that it is very hard to break down into small pieces if it has once been waterlogged and is then dried, thus making the pretreatment procedure more difficult. However, some laboratories may not mind this.

Remember that a laboratory will look more kindly on samples if they are not accompanied by a ton of soil. A certain amount of physical precleaning of samples can be done before they reach the radiocarbon laboratory, such as concentrating charcoal by extracting it from earth using metal tweezers, but if it is a widely disseminated sample, the first question is whether it is even worth dating. Perhaps most important to check with a laboratory that is likely to be dating the samples is what they would prefer for each type of sample.

Using radiocarbon results

Rarely is the interpretation of radiocarbon results completely straightforward. Occasionally a sample is dated simply to determine roughly whether an object is modern or of considerable antiquity; in essence, an authenticity test. Even then the answer may not be clear cut if, say, an old timber has been recently carved to produce an authentic-looking sculpture! In archaeology, the questions are often quite complex, involving non-contemporary samples. The difficulty arises from the necessity to calibrate radiocarbon results and the form of the calibration curve, precluding both the use of normal statistical tests to answer such questions and the use, other than in the broadest sense, of radiocarbon dating as a relative dating method.

Radiocarbon and relative dating

Prior to an agreement on which calibration curve to use, many archaeologists took the pragmatic approach of working in uncalibrated radiocarbon results rather than calibrate only to find that recalibration was necessary the next time a new curve was produced. This approach has, however, led some users of radiocarbon results to hold a rather spurious belief in a radiocarbon timescale that can be used as a relative dating technique. Unfortunately, this is only the case in a rather limited sense. There are several periods in the calibration curve where events that are separated in calendar time by several centuries appear contemporaneous from their radiocarbon results. The worst of these (see fig. 23) is for the period corresponding to the British Early Iron Age (c. 800–400 BC). There are also periods where the curve is steep, so that an apparently large difference in radiocarbon results arises from events separated by relatively small amounts of real time. There could even appear to be an inversion of events if the calibration curve is particularly wiggly and the error on the results is sufficiently low (fig. 25). Radiocarbon results can also appear

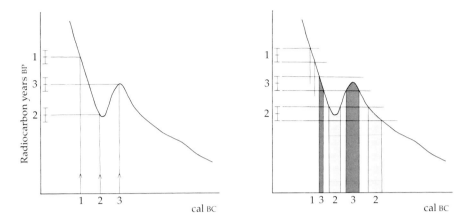

25 If radiocarbon results are used for relative dating, they can falsely suggest approximate contemporaneity of non-contemporary events and vice versa; even apparent inversion of events is possible. In the first diagram a perfectly feasible, if perhaps infrequent, situation is illustrated. Three events, equally spaced in calendar time, are radiocarbon dated with reasonable precision (say, ±40 years). Due to the wiggly nature of the relevant portion of the calibration curve, these events appear in a different order on the radiocarbon timescale. Only when the radiocarbon results are calibrated, as shown in the second diagram, is it apparent that confusion over the order of events 2 and 3 is possible (for the sake of clarity, the calibration of the result for event 1 is not shown). If these events have an archaeologically unequivocal stratigraphic relationship, then it would be possible to eliminate some of the calendar ranges. Such a stratigraphic relationship would also have demonstrated, even prior to calibration, that radiocarbon results do not necessarily offer relative dating, particularly for sequences covering a short time span.

to bunch around a temporal hiatus. The radiocarbon timescale continually compresses or stretches real time so that great care has to be exercised in using it for relative dating, particularly over a timescale of only a few centuries.

Combining results

Replicate measurements
If more than one radiocarbon measurement is made on a single sample, these replicate results can usually be combined. Of course, the sample itself should not represent an age span; if it does, then the various measurements will only be true replicates if the same age range is measured, for example, taking sections of the same tree rings from a large timber.

To quantify possible non-consistency of the results, a chi-square test can be done. This tests whether or not the variability of the results amongst themselves is consistent with the individual quoted error terms. If the variability is substantially larger, then, assuming the errors are correct, the results are not consistent with dating of a single sample and it is not valid to combine them. This could happen if the measurements were on different chemical fractions of a contaminated sample and then perhaps, strictly speaking, they should not be considered as replicates.

The test statistic is

$$\chi^2 = \sum \frac{(t_i - t)^2}{\sigma_i^2}$$

where t is the pooled mean of the individual radiocarbon results t_i and is given below, and the symbol Σ denotes summation of all the terms. The calculated value of χ^2 is looked up in a set of chi-square tables to determine if the variability is too great to be attributed to chance, given the errors involved. If the test is passed, in other words if the results conform to a normal distribution potentially representing a single radiocarbon 'age', then the results can be combined.

The formula for combining a number (n) of replicate results (t_1, t_2, ..., t_i, ..., t_n) is

$$t = \frac{\Sigma\, t_i/\sigma_i^2}{\Sigma\, 1/\sigma_i^2}$$

This formula 'weights' results according to their associated error term σ_i; more weight is placed on results with small error terms than on ones with large errors.

The error σ on the pooled mean is given by

$$\sigma = \sqrt{\left(\frac{1}{\Sigma 1/\sigma_i^2}\right)}$$

In the situation where all the σ_i values are the same, the formula reduces to the familar one for averaging

$$t = \Sigma\, t_i/n$$

and for σ

$$\sigma = \frac{\sigma_i}{\sqrt{n}}$$

Replication should not be undertaken simply to achieve higher precision (see p. 40). Often multiple dating of a sample is done where contamination is suspected and different chemical components are extracted.

Combining results from different laboratories
If different laboratories, whether conventional or AMS, produce a radiocarbon result for parts of the same sample, their results can be combined as outlined above provided none of the laboratories has a systematic bias and each evaluates its error terms in approximately the same way. Significant systematic errors will become evident on applying the chi-square test, as might differences in error evaluation. If there is some doubt about how the laboratories have estimated their quoted errors, the variability of the results relative to each other can be used, without weighting, to provide a mean result, a standard deviation and hence a standard error on the mean. The usual error estimation process (see p. 38) treats the error as if known. However, in this situation, uncertainty in the error evaluation can be taken into account by use of Student's t-distribution, rather than the Gaussian. Where n results are involved, the mean is

$$t = \sum \frac{t_i}{n}$$

The standard deviation is

$$\sigma_{sd} = \sqrt{\left(\frac{(t_i - t)^2}{n-1}\right)}$$

and the error on the mean is

$$\sigma_{se} = \frac{\sigma_{sd}}{\sqrt{n}}$$

This is the error that should be used in conjunction with Student's t-distribution to provide the confidence levels for the true result.

Combining results from different samples
If there are several samples from the same context and this is believed to represent a short-lived episode, then, with certain provisos, the results can be combined as outlined above. It must be remembered that the radiocarbon result for the sample is not the same as the radiocarbon age of the context. Re-use, residuality and age-offsets can all play a part. In combining results, those on large timbers, for example, should not be combined with those on bone from the same context. Usually this will be apparent from the radiocarbon results themselves, with the former being noticeably older than the latter (in fact, this situation should not normally arise, since the bone samples should be dated in preference to large timbers where the choice is available). However, even the results from several bone samples from the same context need to be considered with care; in particular, some samples may be residual. A chi-square test should be done and, if the results do not pass the test, they are not consistent with dating of a single episode and it is not valid to combine them.

If the test is passed, in other words if the results conform to a normal distribution potentially representing a single radiocarbon 'age', it must not be assumed that this proves the samples are from the same age population. Rather this evidence *together with* the archaeological evidence indicates comparability. It is also advisable to consult the calibration curve just to see if it is possible that non-contemporary events in calendar years might give effectively the same radiocarbon results, as for example in the period 800–400 BC (see above and fig. 23). Again, results from different laboratories, whether conventional, AMS or some from each, can be combined in this way if the conditions are satisfied and their errors are fully evaluated.

In any combining of results, reservoir corrections and corresponding adjustment of the error term should be done first, but the error term on the calibration curve does not enter these calculations because the assumption is that the samples would have the same ^{14}C content if in equilibrium with the atmosphere. The error on the curve must be taken into account, however, when the pooled mean result is subsequently calibrated.

Comparison of results for different episodes/events

If results are obtained for different events or are shown to be inconsistent using the chi-square test, what statistical procedures can be done on them? Here the difficulty is that an age difference is indicated, but the true magnitude of that difference cannot be evaluated until the individual radiocarbon results have been converted to calendar years by calibration. Of course, as soon as this is done, the Gaussian probability distribution of the uncalibrated result is replaced by graphical date ranges such as those illustrated in figure 20. Statistical tests cannot then be done. It is not valid to perform the tests first and

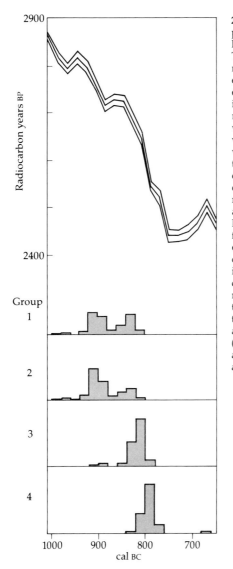

26 Probability calibrations for different phases of the waterfront activity at the Runnymede Late Bronze Age settlement. These are based on twenty-six radiocarbon measurements relating to four discrete events stratified in one deep sequence; one event, the construction of the inner palisade, is represented in fig. 24. Each group of measurements is on material from *in situ* wooden structures or short-lived events, and virtually all measurements are on young wood (less than thirty years' growth), so that there is minimal risk of an age offset between calibrated dates and the true date of the events in question. Within each group the results are consistent both statistically and archaeologically with a single event. Probabilistic calibrations are therefore shown for the mean radiocarbon result for each event. The relevant portion of the calibration curve (illustrated at the top of the diagram) is steep, although a minor wiggle in the curve at 2700 BP leads to bimodal calibrated ranges for groups 1 and 2. It can be seen that the calibrated date ranges correspond well to the sequence observed in the stratigraphy and suggest a time lapse between events 1 (the earliest) and 4 (the latest) of approximately a century and possibly even as little as forty years.

then to calibrate. Nor is it valid to perform the tests and work only in radiocarbon results, because the radiocarbon year is not a true unit of time but is variable in length as previously discussed.

Graphical representations of the calibrated results will help, using either calendar date ranges if the intercept method has been used, or cumulative probability distributions if a probability method has been used (fig. 26). Unfortunately, these do not allow a succinct quantification of the data; for example, phase duration cannot be simply enumerated. However, even without the difficulties caused by calibration, questions such as phase duration involve problems that are inherent to some degree in all sampling of archaeological sites. The underlying assumptions are that the radiocarbon samples selected for dating are representative of the chronology of the archaeological record (for example, they are not biased to earlier or later periods) and that in turn the archaeological record is representative of past human activity; each inference may in reality be a considerable leap.

Rejecting radiocarbon results

In the datelists published in the journal *Radiocarbon*, submitters provide a brief comment on how the radiocarbon results compare with the archaeology and therefore with expectation. Comments such as 'archaeologically acceptable', while not very informative, are less frustrating than the bald 'archaeologically unacceptable' statements. Often there is no discussion of these 'unacceptable' results; they are simply rejected by the archaeologist when evaluating the chronology of the site. Such unexpected or anomalous results can, however, be of great value. For example, they might alert the user to a problem with the laboratory (or vice versa!). Alternatively, they might indicate one of a multitude of depositional problems, such as that the samples selected were residual or that there was unsuspected contamination. These 'unacceptable' results, perhaps more than any others, need careful consideration: they may provide the greatest information.

Sampling strategy

Radiocarbon dating anything and everything, just because it is there and because it is organic, is not a sampling strategy! The literature abounds with results that are of little or no use to archaeology as a result of this 'policy'. Some of the problems of radiocarbon dating and how archaeological depositional processes might affect selection of samples for dating have already been discussed. In summary, any strategy should:

- Involve the radiocarbon laboratory at an early stage.
- Ask how the context yielding the sample relates to the event that is to be dated, how the context was formed and what it means.
- Ask how the sample relates to a given context: is there good association, is the sample representative, is its deposition contemporary within reasonably narrow limits with the context?
- Ask if the ^{14}C activity of the sample is relatable to the time of death of the plant or animal from which it is derived, or whether there is an age offset and, if so, if it is acceptable.
- Ask if the contexts being sampled adequately represent the human activity that is being studied.
- Ask whether radiocarbon results after calibration can provide the resolution needed to answer the archaeological questions being posed.

Used well, radiocarbon is a very powerful and widely applicable technique, invaluable to our understanding of the unwritten past.

Further Reading

Background reading and general texts

Baillie, M.G.L. *Tree-ring Dating and Archaeology.* Croom Helm, London, 1982.

Eckstein, D. *Dendrochronological Dating.* European Science Foundation Handbooks for Archaeologists No. 2, Strasbourg, 1984.

Gillespie, R. *Radiocarbon User's Handbook.* Oxford, 1984.

Libby, W.F. *Radiocarbon Dating.* Chicago University Press, 1952, 2nd edn 1955.

Mook, W.G. and Waterbolk, H.T. *Radiocarbon Dating.* European Science Foundation Handbooks for Archaeologists No. 3, Strasbourg, 1985.

Taylor, R.E. *Radiocarbon Dating: An Archaeological Perspective.* Academic Press, London, 1987.

Waterbolk, H.T. 'Working with radiocarbon dates', *Proceedings of the Prehistoric Society,* 1971, v. 37, pp. 15–33.

Conferences

The radiocarbon community holds an international conference every three years, the proceedings of which are published in the journal *Radiocarbon*. The coverage is wide, from technical problems and advances to applications in various fields including archaeology. The AMS laboratories also hold a triannual specialist conference, and there is now a regular meeting for archaeologists and radiocarbon scientists held in Groningen in the Netherlands. The proceedings of the first Groningen conference are published by *PACT*, the journal of the European Study Group on Physical, Chemical and Mathematical Techniques Applied to Archaeology. However, at the time of writing, the proceedings of the second meeting held in 1987 are still unpublished. In addition, 'one-off' meetings with a specific purpose may be published. Notable of these in the recent past are:

Gowlett, J.A.J. and Hedges, R.E.M. (eds). *Archaeological results from accelerator dating.* Oxford University Committee for Archaeology Monograph 11, 1986.

Ottaway, B.S. (ed). *Archaeology, dendrochronology and the radiocarbon calibration curve.* University of Edinburgh, Department of Archaeology Occasional Paper 9, 1983.

Finding dates

There are probably over 200 radiocarbon laboratories worldwide, of which about 130 are listed at the back of each volume of the journal *Radiocarbon*. However, few are dedicated to dating only archaeological samples. In theory, all laboratories publish their results as datelists in *Radiocarbon*, but in practice only a small proportion do. Alternatively, results for archaeological materials may be found in excavation reports and occasionally in scientific journals, notably those dedicated to science applied to archaeology (in particular, the dates from the Oxford AMS laboratory are published in *Archaeometry*). Hence finding the information you want, even if it is published, can be a process of trial-and-error. To alleviate this problem, computerised databases are now being established, the most comprehensive of which is being co-ordinated by Renee Kra, one of the editors of *Radiocarbon*.

Selected bibliography relating to the illustrations

Aitchison, T.C. *et al.* 'A comparison of methods used for the calibration of radiocarbon dates'. *Radiocarbon*, in press.

Aitken, M.J., Michael, H.N., Betancourt, P.P. and Warren, P.M. 'The Thera eruption: continuing discussion of the dating'. *Archaeometry*, 1988, v. 30, pp. 165–82. (*See fig.* 9)

Burleigh, R. *et al.* 'British Museum natural radiocarbon measurements X'. *Radiocarbon*, 1979, v. 21, pp. 41–7. (Dates for Grimes Graves; *see fig.* 2)

Coles, J.M., Orme, B.J. and Jones, R.A. 'Withy Bed Copse, 1974'. *Somerset Levels Papers*, 1975, no. 1, pp. 29–40. (*See fig.* 22)

Damon, P. E. *et al.* 'Radiocarbon dating of the Shroud of Turin'. *Nature*, 1989, v. 337, pp. 611–15. (*See fig.* 13)

Gowlett, J.A.J., Hedges, R.E.M. and Law, I.A. 'Radiocarbon accelerator dating of Lindow Man'. *Antiquity*, 1989, v. 63, pp. 71–9. (*See fig.* 21)

Levin, I. *et al.* '25 years of tropospheric ^{14}C observations in central Europe'. *Radiocarbon*, 1985, v. 27, pp. 1–19.

Needham, S.P. *Excavation and salvage at Runnymede Bridge: the Late Bronze Age waterfront site.* British Museum Publications, London, in press. (*See figs.* 24 & 26)

Pearson, G.W. 'How to cope with calibration'. *Antiquity*, 1987, v. 231, pp. 98–103.

Sieveking, G. de G. 'The Kendrick's Cave mandible'. *British Museum Quarterly*, 1971, v. 35, pp. 230–50. (See *Archaeometry*, 1985, v. 27(2), for publication of the date; *see fig.* 14)

Stead, I.M., Bourke, J.B. and Brothwell, D. *Lindow Man: The body in the bog.* British Museum Publications, London, 1986. (*See fig.* 21)

Wilson, I.W. *The Turin Shroud.* Victor Gollancz, London, 1978. (*See fig.* 13)

Index